Data Monitoring Committees (DMCs)

Nand Kishore Rawat • David Kerr

Editors

Data Monitoring Committees (DMCs)

Past, Present, and Future

Springer

Editors
Nand Kishore Rawat
Lantheus
King of Prussia, PA, USA

David Kerr
Seattle, WA, USA

Editorial Contact
Carolyn Spence

ISBN 978-3-031-28762-6   ISBN 978-3-031-28760-2   (eBook)
https://doi.org/10.1007/978-3-031-28760-2

To the patients who participate in clinical trials and the entire clinical research community who make it happen

Preface

This book provides an overview of Data Monitoring Committees – what was done in the past, what is currently being done, and thoughts on improvements for the future.

Virtually everyone reading this book will take a medicine or treatment at some point in their lives that was first studied in a clinical study – many of which had Data Monitoring Committee oversight as a key agent to protect patient safety. But virtually nobody has had any formal training on DMCs. Neither the people working at the companies developing the medicine or treatment, nor the prospective DMC members. This book's goal is to educate all those involved in the DMC process on the best practices for DMC.

The authors have attended ~1000 DMC meetings from ~250 distinct studies across all areas of clinical studies (oncology, rheumatology, rare diseases, cardiology, immunology, etc.). This wide range of experience helped shape the creation of this book, as well as their experience that comes from working on DMCs with virtually every large biotechnology and pharmaceutical company.

The reader of the book will learn when DMCs are needed or helpful, how to form the DMC, how to work with external CROs and with sponsor teams and the DMC to create needed DMC outputs, how the DMC meetings are conducted, and – especially for DMC members – what are considerations within the Closed Session to review for risk/benefit to make appropriate recommendations that protect the patient safety and trial integrity.

The book provides the DMC framework for biotechnology and pharmaceutical professionals across the board – current and prospective DMC members, employees in industry and academia and non-profit who are running clinical trials that need DMC support, and employees at CROs that are directly facilitating DMC services.

The topic of DMCs is not generally taught in school, and people might work in industry or academia on clinical trials for decades but not be exposed to DMCs but then suddenly be assigned to assist on DMC work. So the need was great for a book for people suddenly needing to support a DMC – either at a sponsor, or CRO, or actual DMC member. Previous books have focused on large cardiology studies (which admittedly were the most common studies to use DMCs decades ago), but there was a need for a book that discusses more current studies, along more detailed look on the behind-the-scenes work from the SDAC. We hope this book provides help in navigating the DMC process so that patient safety and study integrity are well protected.

Chapter "Introduction" provides an overview of the types of clinical trials that are conducted, focusing on the values of a randomized, blinded controlled study, and setting the stage for the value of a DMC in such a study.

Chapter "What Is a DMC" provides the high-level overview of the key responsibilities to the study.

Chapter "Is a DMC Required? What Other Groups Are Involved?" discusses when a DMC is recommended to be used, and reviews the other groups involved with the clinical trial that perform complementary activities to help the DMC.

Chapter "Who Is on the DMC?" lists the qualifications for DMC members as well as quorum, voting, and duration of service.

Chapter "What Are the Legal and Financial Aspects of a DMC?" covers legal aspects, primarily regarding conflict of interest since DMC members should be independent oversight.

Chapter "How Does the DMC Work with SDAC and Sponsor and External Groups?" delineates the key interactions between the DMC, the sponsor of the study, and the group facilitating the DMC activities.

Chapter "What Does a DMC Meeting Look Like?" talks about the different types of DMC meetings, and what types of sessions and discussion will exist for each of these types of DMC meetings.

Chapter "What Data Is Used for DMC Outputs and Who Programs?" focuses on the data that is used for the DMC review and who is doing the programming behind the scenes for the outputs the DMC receives.

Chapter "What Is Included in DMC Outputs?" is a general overview of these DMC outputs, clearly emphasizing that the layout and table of contents for outputs created for a DMC are expected to be very different from what would be generated at the end of the study for regulatory review.

Chapter "What Do the Final DMC Outputs Look Like and How Is It Delivered?" continues along those lines for how final materials are packaged and distributed to the DMC.

Chapters "What Types of Safety Outputs Does the DMC Receive?", "What Types of Efficacy Outputs Does the DMC Receive?" and "What Types of Other Outputs Does the DMC Receive?" go into detail about the outputs that are both comprehensive and comprehensible for the DMC – safety, efficacy and other outputs respectively in the three chapters.

Chapter "What About In-Between DMC Meetings?" mentions materials that might be sent to the DMC outside of the normal DMC meetings.

Chapter "What Types of Formal Interim Analyses Does the DMC Review?" is an in-depth discussion of formal interim analyses with examples and only gently touching on advanced statistical theory.

Chapter "What Does the Paperwork from DMC Meetings Look Like?" gives guidance on the formal documentation that should be in place after each DMC meeting takes place.

Chapter "How Does the DMC Assess Risk-Benefit for Their Decision Making?" focuses on the DMC decision making, particularly when faced with difficult choices.

And, finally, chapter "What Are Some Examples?" provides a variety of vignettes culled from the author's experiences.

Writing a book is harder than we thought and more rewarding than we could have ever imagined. None of this would have been possible without help and support of our colleague Bill Coar, who assisted with SAS code to produce many of the figures included in this book. Data for some figures was obtained from pilot data from CDISC (Clinical Data Interchange Standard Consortium) and appreciation is made to all those involved in CDISC's valuable activities in clinical trials.

We're thankful to Lingyun and Cyrus who agreed to write the Interim Analysis chapter for the book. This chapter covers an in-depth discussion of formal interim analyses with examples and only gently touching on advanced statistical theory.

We're thankful to our industry friend, Rich Davies, who heads the safety statistics division at GSK. He assisted with one of the case studies.

David appreciates the hundreds of collaborative, diligent, and knowledgeable co-workers he has had the good fortune to work with (and become friends with) over the past 25 years at what started as SERC, which became Axio Research, and then part of Cytel. In particular he thanks Ruth McBride who first hired him as an eager but naïve statistician and introduced him to the world of biostatistics and DMCs, and he thanks Lee Hooks who taught him the business side of CROs, and he thanks Kent Koprowicz for always looking for ways to improve the DMC process (and for the past five years saying, "You should write a book!"). Especially thanks to all who played lunchtime card games – especially the bridge group of Lee, Kent, and Brian Ingersoll. Appreciation to the hundreds of DMC members David has learned from when facilitating DMC meetings, and the dedicated teams at sponsoring companies striving to prevent and cure disease.

David's wife, Merissa, suggested not to include her in the acknowledgments since she claims she didn't even notice him writing this book. But on this rare occasion he will disregard her opinion and thank her for all the support managing the household while he was distracted by work these many years. And David thanks his two children Natalie and Allison for all the fun they've had together – and he thinks they now understand what he does even if in their early years David simply described his work as "meetings and memos". Of course, David could not have done this without the lifelong support from his parents, Les and Arlyn.

Nand thanks his wife, Sprha, for her love and constant support, for all the late nights and early mornings, and for taking care of kids while he was crazy busy "thinking and writing" and keeping him sane over the past year. Most of all, he thanks you for being his best friend. He owes you everything. He is grateful for his beloved daughters, Yashasvi and Yushika for being such a bundle of joy and cheering up for him always. Yashasvi always made sure to check on the status of the book submission deadline while Nand used to put her to sleep at night.

To Nand's family, particularly his parents, some of the most important lessons of his life have come from them, and for that he is forever grateful. To his mother- and father-in-law, who have welcomed him into the fold over the last decade and he truly feels like he is family.

King of Prussia, PA, USA   Nand Kishore Rawat
Seattle, WA, USA   David Kerr

Contents

Introduction

David Kerr and Nand Kishore Rawat

Abstract This chapter provides an overview of the types of clinical trials that are conducted, focusing on the values of a randomized, blinded controlled study, and setting the stage for the value of a Data Monitoring Committee (DMC) in such a study. The DMC provides value by reviewing safety and efficacy data, and then making recommendations. These recommendations are made to help ensure study integrity and to protect the patients – ultimately up to recommending stopping the study to protect patient safety against undue risk, or due to established statistical futility or statistical benefit.

Keywords Clinical study · Experiments · Hypotheses · Controlled study · Blinded study · Randomized study · Randomization · Data Monitoring Committee (DMC)

Some of the greatest advancements in human development have been in the understanding of diseases and how to prevent and cure them. For centuries, scientists and doctors have generated hypotheses and conducted experiments to see if the hypotheses arc indeed borne out by evidence. They might initially start by experimenting on animals (pre-clinical studies), but eventually they will want to test the theories in people (clinical studies). If the results of these clinical studies show the new treatment is beneficial and safe (or at least safe enough, given the level of benefit), then regulatory agencies will approve these new treatments for use and hopefully be embraced by the clinical community.

These clinical trials are typically conducted by comparing two or more treatments to each other. For example, some patients might get the new treatment, whereas others would get the standard of care (the current best practice) or others might get a placebo (a non-active treatment). It's critical that these clinical trials have a control group to compare the new treatment against. Without a control group

D. Kerr
Seattle, WA, USA
e-mail: david.kerr@cytel.com

N. K. Rawat (✉)
Lantheus, King of Prussia, PA, USA

© The Author(s), under exclusive license to Springer Nature
Switzerland AG 2023
N. K. Rawat, D. Kerr (eds.), *Data Monitoring Committees (DMCs)*,
https://doi.org/10.1007/978-3-031-28760-2_1

to compare against, there is a great challenge to determine if the new treatment actually produced results better than the standard of care or placebo would have. Comparisons against historical data are generally flawed. Having a controlled study (treatments directly compared against each other) is the standard practice.

Another aspect of clinical trials is that (if possible) they are conducted in double-blind or triple-blind fashion. This means that the subject and treating physician (double-blind) or possibly the sponsoring company running the study as well (triple-blind) do not know whether the patient is taking a new treatment or the comparator (standard of care or placebo). This helps enhance the integrity of the study. If differences in benefit are seen, it makes it more likely that the benefit is due solely to the new treatment, not to differences in how patients or treating physician consciously or subconsciously reacted to the knowledge that the subject is or is not on the new treatment. For these blinded studies, an outside group is responsible for packaging kits, pills, etc. with the intent that all treatments appear to same to the patients and physicians. Not all studies are amenable to having blinded treatments – for example, a comparison of a treatment given intravenously compared to a treatment given subcutaneously might not have 'sham' administration of the other treatment.

A final component of many clinical trials is randomization. A subject entering the study could end up being randomized to receive the new treatment, or the comparator. And if the study is blinded, they would not know what they received. Randomization also helps enhance the integrity of the study. Randomization will help ensure that the types of patients in each group are similar. So if benefit is seen in the new treatment, it's not because there were healthier or sicker patients, or any differences in demographics. In a large enough clinical trial that employs randomization and blinding and a control group, if a benefit is seen in the new treatment compared to the control group, we are reasonably confident that it is because of the new treatment and not some outside influence.

So imagine a controlled, blinded, randomized clinical trial that will enroll 200 patients with acute myeloid leukemia (AML) that will randomize 1:1 – all patients will get standard of care for AML, and all patients will additionally take daily pills, 100 patients will take pills with the new treatment, and 100 patients will take placebo pills, and no one in or involved directly with the study will know who is taking the pills with the new treatment and who is taking the placebo pills. Our study will take 2 years to enroll, and we want to follow the subjects to learn about deaths. Our study will be complete after 120 subjects die. Our hypothesis based on previous experience is that of those 120, ≤ 50 will be in the group that received the new treatment and ≥ 70 will be in the group that received placebo. Those results would be statistically and clinically compelling to regulatory agencies and the clinical community. (Statistically compelling because it's a bigger difference in results than would have been expected by chance, and clinically compelling because this difference represents appreciable improvement in patient outcome.) It might take another 2 years after enrollment is complete before the 120th patient dies.

So imagine what might happen at the end of the 4 years, remembering that the patients, treating physician, and company sponsoring the study do not know what treatment the individual subjects are on. After 4 years, the sponsor finalizes the database with all of the subject data, and then (virtually) 'rips open the envelope' that contains the information on what treatment each subject had and analyzes the results.

One possible situation is that the results are amazing. Of the 120 deaths, 40 were on patients who took the new treatment and 80 were on patients that only had the standard of care. This is great news – but then you start to look at the data in more detail and realize that even a year ago, when only 80 deaths had been occurred, the results are also very impressive at 30 vs. 50 deaths. Perhaps those results would have been impressive enough to stop the clinical trial early and convince the regulatory agencies and clinical community and start to deliver this new treatment to prospective patients (and start selling the new treatment) 1 year earlier.

Another possible situation is that the 4 years finish, the virtual envelope is opened, and it is revealed that of the 120 deaths, there were 60 on the new treatment and 60 on the placebo. This is obviously disappointing. After looking closer at the data, it's observed that even earlier on, it was clear that there was no benefit being obtained – a look after 80 deaths a year earlier would have shown a 40 vs. 40 split between the new treatment and placebo. The final year of the study was a waste of resources, and the patients could have started other treatments earlier.

Another possible situation could be that the study finishes, and it's revealed that the new treatment actually had more deaths, 65 vs. 55. And not just more deaths, but substantially more toxicity seen from patients treated with the new treatment – adverse events such as serious infections were much more common on the subjects taking the pills with the new treatment. And in fact this lack of benefit and concerning excess toxicity of serious infections was occurring earlier in the study. Wouldn't it have been good to have detected this and stopped or substantially changed how the study was being conducted to protect patients taking the new treatment from risks that were happening without any likely accompanying benefit.

We noted the value of the triple-blinded study. So how can these three scenarios above be detected if the patient, the treating physician, and even the sponsoring company do not know the results by group until the conclusion of the study. The answer is a Data Monitoring Committee (DMC). This group has no involvement in the day-to-day activities of the study but is tasked to periodically review interim results to ensure the study is still ethical to continue. Sometimes called by other names – e.g., Data Safety Monitoring Board (DSMB) – the use of DMCs in increasing to ensure that patient interests (patients on the study as well and prospective patients) are upheld.

This book will discuss the DMC process in detail – who is on the DMC, how the DMC operates, what type of information the DMC gets to make recommendations, what considerations the DMC makes as they form recommendations. Hopefully, this will help if you are a new or even an experienced DMC member, a member of a sponsoring company that has a DMC overseeing one of your studies, or from an outside organization that is facilitating the DMC's efforts.

And hopefully you find this aspect of clinical trials as engaging as we, the authors, do. Every study has a universe of possibilities before it as it gets started. Watching data from the study evolve in a unique way for each study and observing how DMCs uniquely react to that data mean that there really is no such thing as a 'typical' or 'boring' DMC. Hopefully, this book will be your guide to the philosophy for DMCs and the success of studies and protection of patients.

What Is a DMC?

David Kerr and Nand Kishore Rawat

Abstract This chapter provides the high-level overview of the key responsibilities of the DMC to the study. Additional thoughts are provided about how DMC recommendations are formed – the ethical considerations so that the DMC recommendations can provide benefits to the current patients, the potential patients, the larger patient population, the clinical community as well as the organization sponsoring the study.

Keywords Data Monitoring Committee (DM) · Statistical Data Analysis Center (SDAC) · Recommendations · Safety · Toxicity · Efficacy · Futility · Study integrity · Ethics · Charter · Risk-benefit

A Data Monitoring Committee (DMC) is a group of independent experts who periodically receive (generally) by-arm outputs, created (generally) by an independent Statistical Data Analysis Center (SDAC) using interim data from ongoing study or studies, so that the DMC can make recommendations about the continuation of the study or studies, based on their best judgment, and sometimes specified guidelines.

The DMC will make recommendations to the sponsor, and the ultimate decision on these recommendations will be up to the sponsor.

The DMC's recommendations will (generally) consider aspects of safety, efficacy, and study integrity. Overarching on all of these is ethical considerations. If the new treatment is excessively toxic compared to the comparator, then the DMC may feel ethically compelled to recommend changes to the study conduct – possibly to recommend stopping treatment or enrollment in the study – to protect current and potential participants. And if the new treatment is so overwhelmingly efficacious, then the DMC may feel ethically compelled to not continue treatment and

D. Kerr
Seattle, WA, USA
e-mail: david.kerr@cytel.com

N. K. Rawat (✉)
Lantheus, King of Prussia, PA, USA

© The Author(s), under exclusive license to Springer Nature
Switzerland AG 2023
N. K. Rawat, D. Kerr (eds.), *Data Monitoring Committees (DMCs)*,
https://doi.org/10.1007/978-3-031-28760-2_2

enrollment on an inferior option. And if the new treatment is unlikely to demonstrate expected superiority, the DMC may feel ethically compelled to declare futility and not waste the time and good intentions of current and potential participants. The DMC will also review study integrity – if the study is unlikely to produce an interpretable result or produce a result within a reasonable period of time, then the DMC may feel ethically compelled to recommend action.

The DMC is focused on patients, but there is also value to the sponsor. The DMC assessment of futility can lead to savings of millions of dollars when studies are stopped early when there is virtually no chance of a statistically significant result at the end. The DMC is also another external body to help ensure that study is on track to yield interpretable results in a reasonable period of time – that enrollment and/or event accrual are on track, and that data collection and adherence to the clinical trial protocol (the document that specifies the clinical trial's procedures) track with expectations.

The DMC should primarily consider the ethics of the continued treatment of participants in the study. The DMC will also want to consider the ethics of enrollment of potential participants. But less commonly considered is that the DMC considers the ethics of the entire patient population and clinical community. Consider if a treatment is already approved, but a safety signal emerges during a study. How long should the DMC let the safety signal develop so that, if true, results are convincing to a community that is already widely using the treatment? Or what if efficacy has been demonstrated before the study is complete, but worrisome safety concerns have also emerged but are not yet definitive. Should the DMC continue to recommend the study to continue in order to gain more clarity on the possible safety issues, or recommend the study stop early for efficacy but leaving unanswered questions about safety?

Theoretic examples like the above show the value that an experienced DMC – armed with a clearly written DMC charter that describes the DMC responsibilities and process – and working with an experienced SDAC to provide them with materials will help ensure the ethical oversight of the safety, efficacy, and integrity for the study. Current and potential participants in the study may not be aware of the DMC activities or that a DMC even exists, but they will be the beneficiaries of the DMC's oversight.

Is a DMC Required? What Other Groups Are Involved?

David Kerr and Nand Kishore Rawat

Abstract This chapter discusses when a DMC is recommended to be used, and reviews the other groups involved with the clinical trial that perform complementary activities to help the DMC. Regulatory guidance is provided for when DMC should be considered. The DMC has a special place in review of study oversight. However, many other groups also play roles. This chapter will distinguish the roles of those groups and how they are different from what the DMC does. Differences in DMC process for open-label and single-arm studies (typically Phase 1 and Phase 2 studies) are highlighted, contrasting from randomized blinded controlled studies (typically Phase 3 studies).

Keywords Food and Drug Administration (FDA) · Regulatory agency · Contract Research Organization (CRO) · Principal Investigator (PI) · Institutional Review Board (IRB) · Steering Committee · Event Adjudication Committee (EAC) · Blinded Independent Central Review (BICR) · Safety Assessment Committee (SAC) · Double-blind/triple-blind study · Single-arm study · Open-label study · Phase 1/2/3 study · Confirmatory study · Hypothesis-generating study

The Food and Drug Administration (FDA) guidance (FDA 2006) [1] states that sponsors of clinical trials consider using a DMC when:

- "The study endpoint is such that a highly favorable or unfavorable result, or even a finding of futility, at an interim analysis might ethically require termination of the study before its planned completion;
- There are a priori reasons for a particular safety concern, as, for example, if the procedure for administering the treatment is particularly invasive;

D. Kerr
Seattle, WA, USA
e-mail: david.kerr@cytel.com

N. K. Rawat (✉)
Lantheus, King of Prussia, PA, USA

© The Author(s), under exclusive license to Springer Nature
Switzerland AG 2023
N. K. Rawat, D. Kerr (eds.), *Data Monitoring Committees (DMCs)*,
https://doi.org/10.1007/978-3-031-28760-2_3

- There is prior information suggesting the possibility of serious toxicity with the study treatment;
- The study is being performed in a potentially fragile population such as children, pregnant women or the very elderly, or other vulnerable populations, such as those who are terminally ill or of diminished mental capacity;
- The study is being performed in a population at elevated risk of death or other serious outcomes, even when the study objective addresses a lesser endpoint;
- The study is large, of long duration, and multi-center".

The DMC is typically the only group that will review results summarized by treatment arm (in randomized studies) to provide an assessment of risk-benefit in an ongoing basis throughout the study. The SDAC will provide these materials to the DMC and facilitate their deliberations. And the sponsor of the study is in charge of operational aspects of the study – perhaps with the assistance of vendors typically known as Contract Research Organizations (CROs). The sponsor or CRO will provide a Medical Monitor to review overall safety data from the study. However, there are many other groups involved.

Sites will typically be responsible for recruiting subjects, scheduling visits, collecting data, accurately and quickly entering data into the appropriate database, and following up on requested queries. The site's Principal Investigator (PI) will be responsible for the clinical management of the individual subjects under her or his care.

Institutional Review Boards (IRBs) have a responsibility that initially seems similar to the DMC but is very different in practice. The IRBs will ensure that the site(s) overseen by the IRB conduct research that protects the rights and welfare of the subjects in the study. IRBs may receive data from the study during the course of the study but, importantly, typically would not see by-arm results from studies that are randomized.

Steering committees are groups that comprise study leadership. They may be fully academic or a mix with representatives from the sponsor. Steering committees would not typically be provided detailed by-arm information during the study, although they might receive the top-line DMC recommendations.

Event assessment/adjudication committees are specialized groups – typically for reviewing endpoint data or key safety data. Names include Event Adjudication Committee (EAC) and Blinded Independent Central Review (BICR). EACs are commonly seen in cardiology studies. Examples would be a group of cardiologists reviewing cardiac events and deaths to determine if the event meets the strict definition of MACE (major adverse cardiac event), e.g., cardiovascular death, myocardial infarction, stroke, and hospitalization because of unstable angina. BICRs are commonly seen in oncology studies to evaluate if a patient's disease has progressed using standard definitions.

These EAC and BICR committees typically do not have access to treatment information – which is especially important in an open-label study (study is not blinded) where others involved in the day-to-day activity may know the treatment. These committees can provide a blinded assessment without being consciously or subconsciously impacted by that knowledge. The data arising from these committees can be key to the DMC work, especially for formal interim analyses to judge efficacy and/or futility – and for these reviews, the DMC will require that data from these committees are timely and current. DMCs may assess expected vs. actual metrics of timeliness of these external review committees.

The FDA within the United States and global regulatory agencies outside the United States will not typically be involved with the DMC process during the course of the study. These regulatory agencies may request DMC materials (statistical outputs, meeting minutes) at the conclusion of the study. But only rarely is there regulatory interaction with the DMC during the course of the study. Regulatory agencies and IRBs assume that the DMC is the one group involved that will protect patients during the study by using their complete access to the data, including by-arm results.

Some sponsors will form internal Safety Assessments Committees (SACs) to determine if any important adverse events are occurring at a higher rate than expected across an entire program of studies. Their data may include not just completed data that they have full access to, but also selected data from ongoing (possibly blinded) studies. The DMC, or perhaps just the SDAC working on the study, may provide select data (e.g., by-arm results from a small number of different types of adverse events, either specifically requested or those that meet specified statistical criteria) from ongoing studies.

Most of this book will make the presumption that the study being reviewed by the DMC is a triple-blind (neither the patient nor the personnel at the site nor sponsor knows what treatment the subject is on) randomized study. However, DMCs still have a value for randomized open-label studies or other studies where there is a single arm or no randomization.

For example, in a randomized open-label study, typically the sponsor teams will intentionally not summarize by-arm results internally during the course of the study. The sponsor teams (or, at least, a subset of the sponsor team) are typically aware of the individual assignments (although some datasets which reveal treatment information might be kept confidential from some team members). The DMC still has the role of reviewing by-arm results to provide guidance on the study. If the open-label study has by-arm results known to the full study team, there still have been DMCs enacted. The rationale for having a DMC in place is that there is value in having an independent group assess the by-arm differences. One might accuse a review done only by the sponsor of the by-arm results of minimizing the importance of some excess safety risk on the active arm, for example. Similarly in a single-arm study, the study team likely has full information (although there have been examples where the study team remained firewalled away from the primary endpoint) and the DMC reviews the same safety data as the study team. Again, having the independence of the DMC in assessing the study data is valuable.

Decades ago, DMCs were most commonly employed for very large cardiology studies. But now DMCs appear most frequently in oncology studies. But likely every therapeutic area and every phase of clinical development have used DMCs at this point. DMCs are most commonly seen in Phase 3 studies (large confirmatory trials), but certainly DMCs are used for Phase 2 studies (mid-size hypothesis-generating trials) – including seamless Phase 2/3 studies where the DMC might be a critical party in the transition process from Phase 2 to Phase 3 – and Phase 1 studies (small first-in-human trials). In Phase 1 studies, the DMC may prove to be the independent voice to concur or disagree on dose-limiting toxicity (DLT) assessments that dictate dose escalation or be asked to recommend if the next dose cohort should be initiated. In Phase 2 studies, the DMC might review different dose arms and either follow an algorithm or use best judgment to recommend adding or removing a dose group from future enrollment.

Who Is on the DMC?

David Kerr and Nand Kishore Rawat

Abstract This chapter lists the qualifications for DMC members as well as quorum, voting, and duration of service. The DMC should include membership that is independent of the sponsor with a range of expertise (both clinical and statistical) and a range of experience on previous DMCs. Other considerations for qualifications are presented. The DMC size and duration of service and process for member replacement are discussed as well.

Keywords Clinician/clinical member · Statistician/statistical member · Voting member · Non-voting member · Member replacement · DMC Chair · DMC size · DMC duration · Quorum · Voting · Consensus · Independence · Qualifications

The DMC will include members who, collectively, have experience in the treatment of subjects with the disease or condition under investigation and in the conduct and monitoring of randomized clinical trials.

The majority of the DMC members will likely be clinicians. Among the clinicians, most will be directly involved in the treatment of the disease under investigation. However, clinicians from other areas may also be members. For example, if there is an expected side effect, then clinicians from those disease areas might be members. An example would be that in a study of diabetes there might be concern about the cardiovascular side effect of the new treatment and therefore would include both endocrinologists and cardiologists. Similarly, there might be concern about the new treatment inhibiting the body's ability to fight off infections, and an infectious disease expert might be invited, or concerns about hepatotoxicity might warrant including a hepatologist. A pharmacokinetic (PK) expert has been seen on a DMC where the DMC had to make a recommendation based on interpretation of

D. Kerr
Seattle, WA, USA
e-mail: david.kerr@cytel.com

N. K. Rawat (✉)
Lantheus, King of Prussia, PA, USA

© The Author(s), under exclusive license to Springer Nature
Switzerland AG 2023
N. K. Rawat, D. Kerr (eds.), *Data Monitoring Committees (DMCs)*,
https://doi.org/10.1007/978-3-031-28760-2_4

that data of levels of drug in the body. The clinicians preferably have previously been investigators on clinical trials and understand the process of clinical trials. The assumption here is that all these clinicians will be receiving the DMC outputs and attending every meeting and participating in discussion. However occasionally, these are consultants (only brought in as needed by the core DMC and perhaps – or perhaps not – receiving the DMC outputs), and occasionally, they are on the DMC (attending all meetings and receiving the DMC outputs) but as non-voting members. The DMC Charter should state if all members are voting members or not.

Some have advocated for ethicists or patient advocates to be represented on the DMC. In truth, their presence on DMCs is extremely rare in current DMCs, and their lack of participation does not seem to have been detrimental.

It is standard that at least one DMC member is a statistician. This is true even if there is no formal interim analysis for efficacy and/or futility and is even generally true if the study is single-arm without any by-arm comparisons. The DMC statistician typically brings the perspective of looking at data at a high level, rather than focusing on individual patients. The best functioning DMCs have all DMC members collaborate – the clinicians and statisticians all bringing their knowledge together. If the study involves a non-standard interim analysis procedure, there is value that the DMC statistician is familiar with it.

Many DMC members work in academia. Some work at non-profits, or hospitals or CROs or are independently employed. Some are retired. As long as there is no conflict of interest (as discussed in chapter "What Are the Legal and Financial Aspects of a DMC?"), it should not matter. One aspect that is important is that the DMC member has the flexibility in their schedules. DMC service is not generally too time-consuming (perhaps taking about 20 h per year). But there can be DMCs that are more intense. And there can situations that need quick response, or to be available for ad hoc meetings with just a week of notice. Potential DMC members who are too busy to support that level of commitment may not be suitable, even if their professional qualifications are otherwise perfect.

The presumption throughout the rest of this book is that DMC members are independent (external) of the sponsor – and that there is minimal and acceptable potential significant conflict of interest. However, there have been DMCs formed that consist wholly of employees from the sponsor (although typically not involved in the day-to-day operations of the study and instructed not to discuss any confidential DMC materials), or a hybrid model that includes DMC members that are both external and internal to the sponsor. These models would typically be seen in Phase 1 or in open-label early Phase 2 studies. A DMC composed of purely independent (external) DMC members is preferred, but the context of the study may allow some or all of DMC members to be sponsor employees.

Most of the DMC members should have served on DMCs previously. But not all need to (or should) have been on many DMCs. There is value in having a range of expertise with DMCs. There will always be a need for DMC members, and it's in the best interests of everyone involved with clinical trials to make sure that new members are welcomed so that they gain experience and can play the part of senior DMC members once the current generation retires from participation in DMCs. At the moment, there is no well-established central certification program or repository of 'qualified' DMC members. Sponsors and SDACs typically look for potential members for a new DMC by working with the same people they have previously or hearing about potential members by word of mouth from asking colleagues within and outside their organizations. But encouragement should be made to bring new clinicians and statisticians in to become DMC members – especially if they represent previously under-represented minority groups.

It is worth considering bringing on a non-voting DMC member who has never served on a DMC to observe and perhaps participate as a voting DMC member for the subsequent studies from the sponsor. This member would receive all the same materials as the other DMC members but would not necessarily be paid and not actively participate in the DMC discussions. Training programs and courses have been considered for new and novice DMC members, but there is no formal certificate program at this time.

In global trials, it would be reasonable to attempt to get DMC members who at minimum are not all from the same country or geographic region. Getting a wider global representation will help the DMC as a whole appreciate what the standard of care and other considerations would be for those patient populations under investigation. It's presumptuous to have a DMC fully based in the United States, for example, for a study that has a relatively large percentage of subjects enrolled in Europe or Asia. This may make scheduling of meetings more challenging, but the benefits likely exceed the logistical issues.

One DMC member will be designated as the Chair. This member will help lead discussion (keeping discussion focused) and form consensus and make sure all voices are heard. The DMC Chair will sign the recommendations and minutes after the meeting. The DMC Chair may be involved in other activities as specified in the DMC Charter (e.g., review periodic safety events and decide if any further wider dissemination to the full DMC is needed). The DMC Chair can be the DMC Statistician. The DMC Chair would typically be the member who has served on the most DMCs previously.

The number of DMC members is variable, depending on the complexity of the study (or studies, if program-wide DMC), side effects, etc. The minimum number is 3. DMCs have gone up to eight members, in the experience of the authors. The complexity of scheduling meetings increases with the number of members, and the ability of each to actively participate in discussion similarly decreases.

Quorum will be based partially on the number of DMC members. Of course, it is preferable that all DMC members attend. But as the number increases, there is an increased chance that one or more members are not able to attend – sometimes known in advance and sometimes only known at the time of the meeting. A general approach for quorum would be to divide the number of members by two and round to the next highest number – for example, for 3 DMC members quorum would be 2, for 4 quorum would be 3, for 5 quorum would be 3, for 6 quorum would be 4. Typically, the DMC Chair has to attend to form quorum. The DMC statistician might have to attend to form quorum, especially if the meeting is for a formal interim analysis. That being said, if a meeting is about to take place and the DMC

Chair suddenly become unavailable, but sufficient numbers of DMC members are available, there have been examples where the meeting continues with an 'Acting DMC Chair', with subsequent follow-up done with the DMC Chair. This approach better serves the clinical study rather than delaying the meeting for weeks or a month while trying to reschedule for a meeting time that works for everyone.

The DMC Charter may discuss voting, but a well-functioning DMC would rarely need to actively vote. The DMC Chair hopefully can find a consensus that all DMC members can agree to. However, voting rules are put into place in the DMC Charter in case that is not possible. Some have advocated that there should not be an even number of DMC members to prevent split votes. That generally is not an issue and should not prevent any DMCs from being an even number of members. The DMC Charter can specify that the DMC Chair has the deciding vote in the event of a split vote. The voting might be different depending on the scenario. For example, two out of four might be needed to recommend stopping for safety concerns, but three out of four might be needed to recommend to stop for overwhelming benefit.

If a DMC member can no longer continue, the Sponsor is responsible for selecting his or her replacement. A DMC member will sometimes resign independently – for example, they have new potential conflict of interest, they become too busy to serve, they have a fundamental disagreement with the sponsor on how the DMC activities will be conducted. Sometimes, the sponsor will ask (or force) a DMC member to resign – they have missed multiple meetings or been nonresponsive, for example. Occasionally, a DMC member will pass away during the course of the study. Depending on the size of the DMC membership and stage of the study, there could be a need for a replacement or not. The DMC Charter will generally have a section about the process for DMC member replacement.

The duration of the DMC service should be stated in the DMC Charter. For a typical double-blind study, the DMC will serve until the study is locked and unblinded – although the DMC might still informally be involved and discuss the top-line results with the sponsor after that time. Studies that have co-primary endpoints at different timepoints or that have open-label extensions require more thought on when the DMC duration is complete.

For example, in an oncology study with co-primary endpoints where final results in progression-free survival (PFS) will happen earlier (due to quicker accrual of events and larger expected treatment effect) than final overall survival (OS) results, there can be discussion about how widely exposed the final PFS results will be within the sponsor, and how to maintain study integrity as the study continues on to collect information on the needed number of deaths. In some situations, the PFS results will be kept to a small number of people at the sponsor, and the DMC will remain in service as the study continues to accrue deaths. Or perhaps to protect study integrity nobody at the sponsor would get access to PFS results and, instead, the DMC would quietly provide the PFS results to agreed-upon personnel at regulatory authorities. A variety of approaches have been seen on whether and how the DMC continues operating once the necessary number of PFS events have been accrued.

And in a study where subjects move to open-label treatment after a period of time on randomized treatment, there should be discussion whether the DMC continues to monitor the study once all subjects have transitioned from randomized treatment to the open-label treatment, or whether the continued monitoring of safety for these patients can be conducted within the Sponsor.

What Are the Legal and Financial Aspects of a DMC?

David Kerr and Nand Kishore Rawat

Abstract This chapter covers legal aspects, primarily regarding conflict of interest since DMC members should be independent oversight. The start-up activities leading up to having a fully contracted DMC member are discussed. Indemnification is reviewed to ensure DMC members are indemnified by the sponsor, and not the reverse. A detailed review of the different types of conflict of interest (not just financial) is provided. Approaches on DMC member payment are also shown.

Keywords Non-disclosure Agreement (NDA) · Debarment · Contract · Indemnification · Conflict of Interest (CoI) · Independence · Payment

A typical process after identifying potential members is to reach out to potential members and have a Non-Disclosure Agreement (NDA) completed. Once that is done, additional information on the specific study and DMC scope can be provided. If the potential member is interested, contracting efforts can begin. A debarment check of the clinical members is typically conducted. Clinical members do not necessarily have to have an active medical license or be seeing patients (especially if retired). But one would not want a debarred clinical member serving on the DMC.

A contract is put into place with each DMC member. Most commonly, the contract of the DMC member is with the sponsor company. Sometimes, the contract is between the DMC member and the SDAC. Generally, the contract should be with the organization that is coordinating the payment. The DMC member's might sign off on the contract as an individual (perhaps as a one-person LLC or similar), or the DMC member might coordinate the contract with the DMC member's institution, sending for review by the institution's legal department. There can be delays when a DMC member's institution gets involved. The goal would be that the full DMC is

D. Kerr
Seattle, WA, USA
e-mail: david.kerr@cytel.com

N. K. Rawat (✉)
Lantheus Medical Imaging (United States), King of Prussia, PA, USA

N. K. Rawat, D. Kerr (eds.), *Data Monitoring Committees (DMCs)*,
https://doi.org/10.1007/978-3-031-28760-2_5

fully contracted well in advance of the first patient being enrolled so that the DMC Kick-Off meeting and finalization of the DMC Charter can also take place prior to the first patient being enrolled. The language in DMC member contract should not be the usual ones used for site investigators and others who are helping to develop the product. The contracts should clearly note the DMC is an independent scientist not beholden to the interests of the sponsor, but rather to the patients and clinical community.

DMC members are encouraged to ensure that language is put into their contracts that have the DMC member indemnified by the company and not vice versa. There are not many cases of DMC members having legal action brought, but it is easy to imagine situations where shareholders or families of patients would have expected a different DMC recommendation and bring legal action against the DMC. In such a case, the DMC member should expect to have access to the legal resources of the sponsor, rather than needing to pay for lawyers out of his or her own pocket. Sample language for DMC member contracts is provided in a paper by DeMets et al. (2004) [2].

A key aspect of the contract effort is assessment of Conflict of Interest (CoI). This is most commonly thought of as financial but can expand into many other domains. It is critical to determine if there is potential significant CoI, not only at the contracting stage, but periodically (e.g., annually, or at the time of each meeting) throughout the DMC duration. The DMC members must be considered as independent assessors of the data. Those with significant CoI should not serve as DMC members. Generally, it is up to the DMC members (potential or current) to self-report potential significant CoI. The sponsor and/or fellow DMC member will assess the disclosure to determine if the potential CoI would impede objectivity and thus preclude membership on the DMC. The DMC member contract and DMC Charter likely will include language about identifying and disclosing new potential significant CoI. This independence should be both the actual independence of the DMC member, but also the appearance of independence. A reasonable and informed third party who has knowledge of the relevant information, including the safeguards applied, should reasonably be able to conclude that the integrity, objectivity, and professional skepticism of the DMC member are intact.

CoI is not just from personal financial review. It could be financial – but for the DMCs institution. It could be emotional – the DMC member has close friends or relatives at the sponsor or has close friends or family with the condition under review. It could be intellectual – the DMC member has spent their career advocating for one therapeutic approach, and this new study could discredit that approach or is advocating a different therapeutic approach than is being studied. The potential threats to independence, as adapted from resources provided by American Institute of Certified Public Accountants (AICPA 2022) [3], include:

Adverse interest – DMC member interest opposed to sponsor interest.
Advocacy – DMC will unduly promote sponsor interest.
Familiarity – DMC member has long or close relationship with sponsor.
Management participation – DMC member has role in management of study.
Self-interest – DMC member could benefit from sponsor success (or failure).

Self-review – DMC member will not properly assess or disclose conflicts.
Undue influence – DMC member will not use judgment due to threats or promises from sponsor.

Potential DMC members earn that qualification by being experienced in clinical trials and having served on other DMCs and served in leadership positions on other clinical trials and working with other sponsors or perhaps even the sponsor of the clinical trial previously. These in themselves do not necessarily represent significant CoI. Previous consulting work with the sponsor in a limited fashion or ongoing DMCs with the sponsor or competitors likely does not represent significant CoI either.

When evaluating CoI, think about the likelihood of compromising independence, the extent to which this would be detectable, and the impact on the subjects and results. Some significant CoI would include having the DMC member or a spouse or other close relative work at the sponsor or a close competitor or having a leadership role on the study or a closely related study from the sponsor or a close competitor. Generally being on the DMC from other studies from the sponsor or a close competitor is not an issue – although the DMC member must remember to maintain confidentiality of results to only the specific DMCs reviewing each study. Substantial consulting efforts with the sponsor or a close competitor on studies in the therapeutic area once the study has begun likely would be considered significant CoI. It can be awkward if the DMC member's institution is an active site even if the DMC member is not the site investigator – there are financial considerations (the DMC member might be accused of having the study go longer since a longer study brings in more revenue to the institution) and there is an issue that the DMC member might be asked to serve as the back-up to treat enrolled subjects at the site, after having been exposed to interim results.

DMC members do not serve pro bono, as they generally are paid 'fair market rates'. Payment should be sufficient to encourage qualified DMC members to serve, but not so much to appear influential. Different members may be paid different amounts – based on being the Chair, or previous DMC experience, or other factors. The decision will need to be made early on whether to pay a flat rate per meeting or pay hourly. Flat-rate payment is easier to handle logistically, although there may be discussion needed on how to handle ad hoc effort during the course of the study that doesn't naturally fall into a meeting payment. And having the same payment regardless of amount of time spent could be seen as encouraging cursory review. Conversely, having an hourly payment adds an extra level of logistics, and it might – for better or worse – encourage excessive review of the materials and correspondingly high number of hours billed. In either case – flat rate or hourly – provisions will need to be made on how to bill for ad hoc meetings, particularly those that are 'invisible' to the sponsor. A typical expectation would be that the time spent for a data review meeting would be about 5–10 h – about 2–4 h in advance to review previous minutes and the open and closed materials, about 2–4 h at the meeting itself, and then 1–2 h after the meeting to review minutes and any post-meeting follow-up and responses to action items.

How Does the DMC Work with SDAC and Sponsor and External Groups?

David Kerr and Nand Kishore Rawat

Abstract This chapter delineates the key interactions between the DMC, the sponsor of the study, and the group facilitating the DMC activities (SDAC). The responsibilities and interactions for each of the six directed pathways of these three groups are provided in detail. The concept of the Sponsor Liaison is introduced, with more about the specifics of how DMC recommendations are communicated and with follow-up discussion if needed. The reporting statistician from the SDAC also has important responsibilities that are outlined.

Keywords DMC Charter · Sponsor · SDAC · Recommendations · Responsibilities · Stewardship · Independent statistician · Ad hoc DMC requests · Logistics · Open/Closed Minutes · Open/Closed Reports · Open/Closed Meetings · Interactions · Sponsor Liaison

The DMC Charter is a critical document and required for every DMC. It provides the framework for how the DMC will operate and how the sponsor, SDAC, and the DMC will interact. It specifies the charge to the DMC. The Charter should provide proper guidance to the DMC without being overly detailed or restrictive. The DMC Charter can be based on SDAC's template, or a template provided by the sponsor. The sponsor and SDAC should agree to the final draft of the Charter before it is presented to the DMC. The DMC Charter should include guidance to the DMC on what recommendations the sponsor would like to receive. But the DMC Charter should not include details from the protocol that may change via protocol

Figure created by me.

D. Kerr
Seattle, WA, USA
e-mail: david.kerr@cytel.com

N. K. Rawat (✉)
Lantheus, King of Prussia, PA, USA

© The Author(s), under exclusive license to Springer Nature   21
Switzerland AG 2023
N. K. Rawat, D. Kerr (eds.), *Data Monitoring Committees (DMCs)*,
https://doi.org/10.1007/978-3-031-28760-2_6

amendment or names of individual sponsor team members (instead list the role and include the name in an appendix) or restrictions on what recommendations the DMC may make. The DMC Charter is typically signed by all DMC members (at minimum the DMC Chair) and may also be signed by representatives of the SDAC and sponsor. The DMC may include appendices such as a template DMC recommendation form, a list of sponsor team members along with their role and contact information, and proposed table of contents for Open Report and Closed Report. These appendices may be updated without requiring full re-signing of the DMC Charter.

The sponsor has responsibilities:

- Recognize the DMC as being responsible for the stewardship of the trial and being independent.
- Advise and educate the DMC and SDAC on past and present scientific, clinical, and statistical issues concerning the study and new treatment.
- Take responsibility for determining response to external information – for example, protocol amendment, updated Informed Consent Form (ICF) – DMC response could be seen as biased once unblinded to interim data.
- Promptly provide any relevant updates (e.g., amended protocols).
- Promptly respond to DMC recommendations and follow-through on any commitments.

The SDAC has responsibilities:

- Prepare and distribute DMC reports that are useful to the DMC.
- Have independent statistician attend the meetings and present the report to the DMC.
- Have general clinical trials expertise as well as expertise with DMCs and specific knowledge of the study protocol and DMC Charter.
- Have knowledge of the data and the programs used to the create the outputs.
- Provide DMC with technical support and have flexibility to respond to ad hoc DMC requests (perhaps without sponsor knowledge).
- Provide logistical assistance if requested: meeting scheduling, drafting meeting minutes, contracting, and reimbursement.

There is a triangle of responsibility for the DMC functioning smoothly – with the DMC, the sponsor, and the SDAC at the points of the triangle. A simplified diagram is shown in Fig. 1.

SDAC → DMC interactions – The SDAC will provide materials to the DMC as agreed on by the DMC Charter and previous DMC requests. The SDAC may be involved in scheduling efforts and help with meeting logistics and providing draft of the post-meeting documentation to the DMC. The SDAC will assist the DMC with the meeting, providing additional information during the meeting itself of after.

DMC → SDAC interaction – The DMC will request additional material if needed. The DMC will provide comments on post-meeting documentation, and the DMC Chair will provide signature. The DMC should be prompt in replying to queries about scheduling and other meeting logistics.

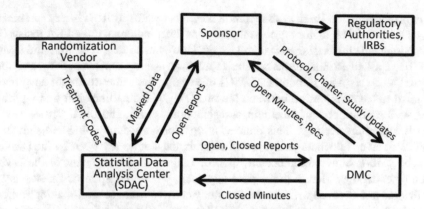

Fig. 1 Communication pathways

Sponsor → SDAC interaction – The sponsor (or designated CRO, as coordinated by the sponsor) is responsible for providing data to the SDAC. Open Sessions of DMC meetings will be primarily led by the sponsor team and slides provided to the SDAC in advance to pass along to the DMC. Confidential data (primarily the randomization data) will be sent to the SDAC appropriately. The sponsor may be aware of that transfer taking place but would not be directly involved.

SDAC → Sponsor interaction – The SDAC (if creating the DMC outputs) will typically provide drafts to the sponsor for review – either with fake randomization or with the open (Total-only) outputs. The SDAC will collaborate with the sponsor on scheduling of meetings. Draft agenda and Open Session minutes will be provided. Timelines for data transfers in advance of each meeting will be discussed.

DMC → Sponsor interaction – There should be very little interaction from the DMC to the sponsor once DMC activities have begun. The final Open Session minutes will go to the sponsor, but usually via the SDAC. DMC members may contact outsourcing personnel with the sponsor after meetings for invoicing activities, if the sponsor holds the DMC member contracts. But it is generally ill-advised to have DMC members interacting with the general sponsor team outside of formal DMC meetings and documentation because of the elevated risk of actual or perceived unblinding or impact on trial integrity. The DMC Chair may occasionally interact directly with the designated Sponsor Liaison (the individual outside of the study team who will receive and respond to DMC recommendations), but many times that action is facilitated by the SDAC.

Sponsor → DMC interaction – There should be no interaction from the sponsor team directly to the DMC after DMC activities have begun. Communication should be done via the SDAC. This removes the chance of a 'reply all' email where a DMC member divulges confidential information, for example. Updates such as protocol amendments, Open Session materials, and enrollment updates can be sent to the DMC via the SDAC.

As seen in the diagram above, it is not the expectation for the DMC to interact outside of the SDAC or Sponsor Liaison. The sponsor (via the Sponsor Liaison) may provide the top-line DMC recommendations to sites or IRBs after each meeting. The sponsor regulatory group may serve as the middleman if the FDA insists on receiving some material the DMC reviewed. However, on rare occasions, the DMC has been asked to talk directly with regulatory agency representatives and has agreed to do so. And sometimes DMCs overseeing the same treatment have been enabled to have discussion between the two DMCs – sometimes just trading minutes with each other after meetings, and sometimes having the DMC Chairs have a quick chat after meetings. This enables a look for consistency of safety signals – similar to an ad hoc meta-analysis of the combined studies reviewed by the DMCs.

The section above references a Sponsor Liaison as the recipient of the DMC recommendations. That is the typical approach, where the Sponsor Liaison is an individual not involved with day-to-day study activity with decision-making authority for the trial or who can forward DMC recommendations to those with decision-making authority. There are variations on this. The Sponsor Liaison might be different for recommendations coming from formal Interim Analysis reviews, as compared to standard data review meetings. The recommendations might go to a senior executive committee within the sponsor (perhaps sent directly to that committee's chair for simplicity). Some studies (most commonly in cardiology) have an executive committee comprised of academic study leadership outside of the sponsor that leads study decision-making. The recommendations in such situations would be sent to the chair of that executive committee. In smaller companies, everyone at the company might be directly involved with the one study under review, and therefore, the Sponsor Liaison would be the Chief Medical Officer (CMO) or Medical Monitor, acknowledging that this individual does still have day-to-day interaction with the study.

What Does a DMC Meeting Look Like?

David Kerr and Nand Kishore Rawat

Abstract This chapter talks about the different types of DMC meetings, and what types of sessions and discussion will exist for each of these types of DMC meetings. This includes DMC kick-off meeting, data review meeting (as well as interim analysis meetings), top-line meetings, and ad hoc meetings. The sessions within a meeting are covered: standard open and closed sessions, recap sessions, and executive sessions. Considerations for meeting format, length, and frequency are provided.

Keywords Open Session · Closed Session · Recap Session · Executive Session · Kick-off meeting/Organizational meeting · Data review meeting · Top-line review meeting · Interim analysis meeting · Silent meeting · Ad hoc meeting · Table, listings, and figures (TLFs) · Action Items · Meeting format · Meeting length · Meeting frequency

There are quite a few reasons that the Data Monitoring Committee (DMC) will meet – with varying attendees and meeting formats based on the purpose of the meeting.

DMC Meetings are usually broken down into several meeting sessions, which can be categorized as either Open or Closed. A typical DMC data review meeting would be an Open Session followed by a Closed Session.

The Open Session will be attended by the DMC, Statistical Data Analysis Center (SDAC), and the study team. In the Open Session, only blinded materials are appropriate for discussion. The purpose of the Open Session is to share any relevant regulatory updates, study status, and topics of interest with the DMC.

D. Kerr
Seattle, WA, USA
e-mail: david.kerr@cytel.com

N. K. Rawat (✉)
Lantheus, King of Prussia, PA, USA

© The Author(s), under exclusive license to Springer Nature
Switzerland AG 2023
N. K. Rawat, D. Kerr (eds.), *Data Monitoring Committees (DMCs)*,
https://doi.org/10.1007/978-3-031-28760-2_7

The Closed Session will be attended by the DMC and the SDAC. Unblinded materials should be discussed in this session. The purpose of the Closed Session is to discuss the by-arm results and determine a recommendation for the study.

Another type of meeting session that can occur is a Recap Session. If this occurs, it is typically after the Closed Session and is an opportunity for the study team (or a subset, or other designated individuals) to quickly hear the DMC's recommendation. There is an appeal to having this immediate response provided to the study team. However, it is generally advised against this session due to the elevated risk of unblinding. In a Recap Session, no unblinded or by-arm discussion should occur beyond the bare necessity contained in the DMC recommendation. The safer approach is to have communications after the Closed Session be in writing with the Sponsor Liaison, although sometimes a one-on-one telephone call is quickly made between the DMC Chair and the Sponsor Liaison to expedite communication of the DMC recommendation.

Sometimes permitted in the DMC Charter, but rarely employed, would be a DMC Executive Session, which would be held without the presence of the SDAC. That would be primarily if the DMC has issues with the SDAC performance. In this case, minutes would have to be taken by the DMC Chair.

The first meeting where the DMC members will meet one another and the sponsor and the SDAC is at the Kick-Off Meeting. After that, there will be periodic data review meetings that occur throughout the study. Finally, it is not uncommon to have a top-line review meeting after the study database has locked. Other meetings that occur during the study are formal interim analyses, silent meeting, and ad hoc meetings.

The Kick-Off Meeting, sometimes referred to as the Organizational Meeting, is the first formal DMC Meeting that the study team, SDAC, and the DMC Members attend together. Only an Open Session will be expected, and no formal recommendation on the continuation of the study will be made. The DMC Charter (although perhaps not yet finalized and signed) will provide general guidance as to when this should occur. Ideally, this Kick-Off Meeting occurs prior to the first subject being enrolled into the study. In theory, there could be events that develop quickly in the first few subjects that would require DMC review and so it would be wise to have the DMC be fully constituted prior to that. The Kick-Off Meeting agenda could include:

- Introduce all key attendees (including roles and responsibilities).
- Discuss the treatment's mechanism of action.
- Review preclinical study results.
- Review previous clinical study results.
- Present the overall study and program design.
- Review known safety risks/Adverse Events of Special Interest (AESIs).
- Provided expected turnaround time of event adjudication/review.
- Discuss what data, if any, is completely confidential from study team (e.g., PK data, biomarkers, adjudication results).
- Review key aspects of the protocol.

- – Inclusion/exclusion criteria.
- – Summary of assessments (including assessments conducted after treatment discontinuation).
- – Statistical analysis plan, including primary and secondary endpoints.
- – Formal interim analyses.
- – Safety management/dose adjustment.
- Review the DMC Charter.
- Review the DMC mock Tables, Listings, and Figures (TLF) shells or, at minimum, the Table of Contents (ToC) of TLFs.
- Allow Question-and-answer between the study team and DMC prior to any unblinding data being shared.

Typically, the DMC Charter will be finalized relatively quickly after the Kick-Off Meeting, incorporating comments made by the DMC at the meeting.

Scheduled data review meetings will account for the majority of DMC Meetings. Sometimes, these are referred to as "Safety Review Meetings," although it is common for non-inferential efficacy data to be included as well so that term artificially implies that only safety data is being reviewed. Scheduled data review meetings should have both an Open and Closed Session scheduled. These meetings will focus on the overall risk-benefit of the study. A formal recommendation will be made at the end of these meetings.

In the Open Session of Data Review meetings, the study status and safety updates will be led by the study team. Topics will typically include:

- Proposed protocol updates.
- Regulatory updates.
- Status of "sister" studies.
- Response to previous action items.
- Short review of important, interesting safety events (with level of detail at DMC's discretion) - treatment assignment will not be included.
- Current and projected enrollment.
- Estimates of key timepoints of the study (last patient in, interim analysis, final endpoint achieved).
- Aspects of protocol adherence (e.g., numbers of important protocol deviation, numbers prematurely discontinued from study follow-up).
- Demographics, baseline disease characteristics, and/or stratification – are these the patients that were anticipated to be enrolled?

DMC Open Tables, Listings, and Figures review is optional. Detailed presentations of the Open TLFs should be avoided unless specifically requested since this is of minimal value to the DMC – much more important is the DMC reviewing these results by-arm in the Closed Session. Questions will be shared, and attendees reminded that both questions and answers should be blinded. A general discussion of the next DMC Meeting date should occur, although detailed discussion will wait for the Closed Session.

Some in the DMC field have strongly advocated against presenting any safety or efficacy data in the Open Session, even when presented in blinded fashion. The fear is that knowledge about the control arm provides information about treatment effect. For example, imagine a study with subjects randomized in equal numbers to new treatment and standard of care. If the overall rate of neutropenia is 30%, and the known rate of neutropenia for the standard of care is 20%, then it's easy to deduce that the rate of neutropenia in the new treatment is 40%. Similarly, if the 1-year death rate for subjects with this disease using information from other sources is 20%, but only 15% of subjects that were randomized a year ago have died so far in the study, one could infer that the 1-year death rate of subjects on the new treatment is only 10%. However, the practical reality is that the study teams already know the total results, and there is so much uncertainty about the true rates in the comparator arms that any guesses made by the study teams are likely to be incorrect. The standard of care evolves, and the population enrolled in this particular study is never quite the same as what was enrolled in previous studies.

In the Closed Session of a Data Review Meeting, the primary purpose is to review the data, focusing on recommendations. These recommendations primarily are based on by-arm comparisons but could also be recommendations made for reasons beyond by-arm comparisons that have to do with study integrity. If there are relevant previous Closed action items, they will be discussed to ensure the response was sufficient or if more follow-up is required. A quick assessment of any new potential Conflict of Interest may be conducted, if specified to do so in the DMC Charter. Closed TLF review and discussion will likely be led by the SDAC if agreed to by the DMC Chair, although the DMC Chair is ultimately responsible for ensuring full review and participation of all attending DMC members. The flow of the review is at the discretion of the DMC and primarily the DMC Chair and depends on maturity of study and level of review done in advance. For a first or second review, or if some of the members did not have time to adequately review in advance, the DMC Chair or SDAC might – briefly – discuss what is included in each output and solicit comments, particularly those that relate to potential safety or study integrity concerns. Once the study is more mature and assuming all members had adequate time to review in advance, the DMC Chair or SDAC might briefly point how previously noted aspects of the report have changed, and note any new, clear imbalances. The SDAC statistician can lead this, but must recognize that he or she is not a voting member and is not a clinical expert – therefore should not editorialize, or over or under state any differences. A useful approach is for the DMC Chair to query all members one by one at the start of the review to solicit points of interest that will be the focus of the review. Once all questions, comments, and findings are discussed, the DMC will agree or vote on their formal recommendation. Given that Action Items are recommendations as well, they should be reviewed at the end of the Closed Session to confirm everyone's understanding – making sure everyone agrees "who will do what, when" for those action items. The next meeting date should then be determined – a specific date or a window of dates or a data-driven timepoint – which may be in line with the Open Session discussion or may be at a different time point which can be noted as an Action Item.

Typically, data review meetings start with an Open Session, followed by a Closed Session. Some in the DMC community advocate for an initial Closed Session, followed by the Open Session and a final Closed Session. This approach allows the DMC to better prepare for the Open Session by giving themselves a chance to discuss among themselves in the initial Closed Session what questions need to be answered by the study team in the Open Session before the DMC embarks on a more thorough review in the final Closed Session. This approach definitely has merit, but is unfortunately only used in a minority of DMCs.

For DMCs overseeing open-label studies where there is complete by-arm information known to the study team (as opposed to open-label studies where a firewall enforces that by-arm information is not available to the study team), there is no distinct report only for the DMC eyes – everyone has access to the by-arm reports. The Open Session is typically longer since the DMC and study team can review the same by-arm report. There should still be a Closed Session of the DMC and SDAC – although it typically is short.

Previously unplanned DMC Meetings or ad hoc meetings can occur for many reasons and focus on different topics such as:

- Safety concerns – the DMC wants to monitor a safety signal closely;
- Fast enrollment – the DMC does not want to wait too long to review safety if it could mean hundreds of individuals being enrolled between meetings;
- Previously unavailable data is now ready – radiology, efficacy, or specialty lab data can be delayed and not available for the initial safety review;
- DMC request – the DMC previously recommended that additional data be provided prior to making a formal recommendation;
- Outside information – Recent published findings from similar clinical trials have caused concern;
- Regulatory request – Regulatory agencies have requested the DMC review specific safety data;
- Study update – The study team would like to discuss a major study update or change with the DMC prior to the next scheduled meeting.

The study team will be made aware of these meetings, and Open and Closed Sessions may or may not be held based on the purpose of the meeting.

Silent meetings are DMC Meetings that the DMC has requested, and that the study team is specifically unaware of. Silent meetings usually will occur because of a risk-benefit concern such as:

- Data included in the standard TLFs caused concern;
- Recent published findings from similar clinical trials have caused concern;
- Regulatory agencies have requested the DMC review the safety data.

The reasons for a Silent Meeting are a subset of the reasons as to why an ad hoc meeting may occur. It is ultimately up to the DMC if they want the study team to be aware of the meeting or not. Only a Closed Session will be scheduled for Silent Meetings. The DMC is always allowed to hold a Silent Meeting if they feel it is in the best interest of the study. Obtaining data for these meetings can be challenging while still trying to maintain secrecy. Having a mechanism set up in advance at the beginning of the study that allows the SDAC to quietly receive data on demand will help preserve the secrecy of the Silent Meeting.

An interim analysis meeting is a predefined formal assessment of either efficacy or futility, which can result in early termination or other major changes to the continuation of the study. An Open and Closed Session should be scheduled. The recommendation options likely differ from the standard recommendation form used in safety review meetings. The Open Session should additionally focus on possible statistical results from the analyses, and how those will translate into suggested recommendations and next steps. (One other option is to have a brief interim analysis Kick-Off Meeting a month before the actual interim analysis meeting – it could be for the full DMC, or perhaps just the DMC Chair and DMC Statistician.) An Interim Analysis Meeting can be combined with a normal data review meeting. It is strongly suggested that safety data be available for the DMC to aid in their assessment of benefit-risk.

Top-Line Results Meetings will occur at the end of a study. Only an Open Session will be expected. Top-line results of primary and possibly secondary endpoints and key safety results will be presented by the study team. This will be held after the study database has been locked and the study team has been unblinded, and no formal recommendation will be made. Usually, these are results and presentations that the study team is preparing for an upcoming conference or summarize results that will go into an upcoming journal article. These meetings allow the DMC to see the final results of the study they have been watching progress over an extended period of time, as well as let the DMC provide unblinded feedback to the study team and explain the context of previous recommendations that they made.

DMC Members and the study team can meet for a DMC Meeting in different formats. The two main formats of DMC Meetings are teleconference and in-person.

Having the Kick-Off meeting in person is valuable to allow for sufficient time to let everyone thoroughly digest the material. Data review meetings, particularly Interim analysis meetings, also can be valuable to hold in-person. A common approach is to try to hold these in-person meetings in conjunction with scientific sessions. However, many times, the DMC members are too busy to attend DMC meetings during scientific sessions, even if they are all in the same city. A more successful strategy is to meet near the airport of a centrally located city. In-person meetings should be held in a neutral location. It would be unwise for appearance's sake to hold the in-person DMC meeting at a luxury location, or at the client's headquarters.

But most DMCs in recent years have met purely by teleconference. The level of confidence in teleconference meetings has risen over the past few years. Nonetheless, there are still issues. Attendees can be more easily distracted by other activities. There is the challenge of communicating non-verbally, although the request can be made that DMC members activate video for the teleconferences. Teleconference meetings could exacerbate the situation of a single loud voice dominating the discussion, or of a hesitancy of a naturally quiet or non-native English speaker to speak up or be understood and appreciated. Technical issues can occur despite everyone's best efforts, caused by an unexpectedly poor Internet connection for example.

The length of each meeting is variable, depending on the complexity and number of studies. Kick-off meetings can last anywhere from 1 to 4 h. A typical data review meeting held by teleconference would have a 30–45-min Open Session followed by a 75–90-min Closed Session. However, data review meetings that cover many studies or have extra complexity could last 3–4 h (1 h Open and 2–3 h Closed). Meetings that are expected to be over 3 h typically should be held in-person. Meetings can be shorter if only an Open or a Closed Session is needed. For teleconference meetings, it is important to be aware of the time zones of different DMC members and key sponsor attendees. Preferably start times would vary so that the member farthest away from a core group of attendees does not always bear the burden of a very early or very late meeting.

The Charter will dictate how often the DMC is expected to meet for data review meetings. This could be milestone based such as: "When 10 subjects have 2 cycles of treatment," "When 50 subjects have 3 months of follow up time," "Six months after the first subject is treated," "Within 1 month of the first death." DMCs may also meet based on a given time interval such as: "The DMC will meet approximately every 4 months," or "The DMC will meet at least once every year, and up to twice a year." Interim analysis meetings are more precisely planned. Most studies should expect to have a DMC data review meeting no more than 6–9 months apart. Meeting more frequently than quarterly is typically too burdensome. Meeting less frequently than every 9 months (even if study activities are slow) hurts because the DMC members lose engagement with the study. The key aspect to balance between calendar time and information time. Rapid enrollment or event accrual would warrant more frequent meetings. Conversely, there is little value to the DMC meeting if there has been little new data accumulated since the previous DMC meeting. It would certainly also make sense to meet more frequently early on for an intervention for which little is known about in this patient population. Once a year or two of data has been collected or if treatment is complete or nearly so and long-term data being collected, the DMC should be able to reduce the frequency of meeting at their discretion based on their knowledge of the safety profile.

If there is a situation that requires high level of frequency (as was seen in DMCs overseeing COVID-19 studies), some review might be done off-line only requiring teleconference if a member requests, or outputs might be abbreviated and quickly reviewed at a standing DMC meeting (say, 2:00 PM ET on the third Thursday of each month).

What Data Is Used for DMC Outputs and Who Programs?

David Kerr and Nand Kishore Rawat

Abstract This chapter focuses on the data that is used for the DMC review and who is doing the programming behind the scenes for the outputs the DMC receives. Data for DMC purposes typically needs to balance currentness vs. cleanliness. The underlying programming of the DMC outputs can be done by the sponsor or its CRO, or the SDAC supporting the DMC. The advantages and disadvantages of these different approaches are listed – noting the importance that in any situation that ad hoc requests by the DMC can be addressed confidentially if needed.

Keywords Clinical cut date · Snapshot date · Data Currentness · Data Cleanliness · Programming approaches

The DMC Report should provide information that is accurate to the extent possible, although not all data will be perfectly clean. Follow-up should be complete, if possible, to the "Clinical Cut Date," which is within 6–9 weeks of the date of the DMC meeting. Strong efforts should be made to have sites have subjects come in for scheduled study visits that take place prior to the Clinical Cut Date, with the expectation that data from those visits will be included in the DMC reports. The database should be provided to the SDAC on the "Snapshot Date" that will occur 3–4 weeks before the DMC meeting. Visits from prior to the Clinical Cut Date will have at least some minimal cleaning done (focusing on adverse events and disposition data). Efficacy or safety events that occur after the Clinical Cut Date but are captured before the Snapshot Date should also be provided and included in the reports, acknowledging that this data will not have had the level of cleaning done for visits and events from prior to the Clinical Cut Date. Promptly after the Snapshot Date, the database should be sent to the SDAC who will be generating the DMC Open and

D. Kerr
Seattle, WA, USA
e-mail: david.kerr@cytel.com

N. K. Rawat (✉)
Lantheus Medical Imaging (United States), King of Prussia, PA, USA

© The Author(s), under exclusive license to Springer Nature Switzerland AG 2023
N. K. Rawat, D. Kerr (eds.), *Data Monitoring Committees (DMCs)*,
https://doi.org/10.1007/978-3-031-28760-2_8

Closed Reports. The DMC should feel empowered to argue against proposals to only give "clean" data to the DMC that excludes any data collected recently and excludes recently enrolled subjects. Excluding this data is detrimental to the DMC's obligation to protect patient safety. The DMC should have the knowledge and tools to understand how to interpret data included that is not perfectly "clean." The discussion above is primarily related to the timing and cleaning of the standard data collected by the study - the case report form (CRF) data. Alternate pathways are almost always needed to transfer the randomization data to the SDAC. Laboratory data might be sent by a separate pathway directly from a central laboratory to the SDAC - particularly laboratory data that is being kept confidential from the study team such as PK data or biomarker data. Examples could be high-sensitivity C-reactive protein (hsCRP) data in a cardiology trial, or lymphocytes in an oncology study where the new treatment is known to be lymphodepleting but the control arm is not. Other pathways might be set up for the SDAC to receive other speciality data, such as data from an adjudication committee or data from a thorough QT study.

It is somewhat common that the first meeting is a subset of the eventual set of materials. It may just be listings, if the number of subjects is very limited. Or it may just focus on key disposition and AE data if the number of subjects is still small. This is especially true if visits after baseline are spaced out and, say, there is very little lab data or other post-baseline visits conducted. This is also the case if data management groups are still finalizing the process for extracting and processing data (e.g., getting adjudication data process initiated, or creation of processed datasets rather than using raw data extracted from the clinical database).

There are a variety of models for how the programing of the DMC can be done. Here are four approaches that have been seen:

1. SDAC receives report from Sponsor (or CRO) and passes directly to DMC. This is most commonly seen in single-arm studies (particularly device studies) and dose-cohort escalation (non-randomized) studies. It is occasionally seen in early Phase 2 open-label randomized studies where the Sponsor is not implementing any firewalls against their own by-arm review.

Advantages:
• Cheapest approach – minimal work needed by SDAC.

Disadvantages:
• Minimal ability by SDAC to assist in ad hoc outputs or understanding of the outputs created.

2. SDAC logs directly into firewalled area of Sponsor environment and creates TLFs after swapping in real randomization and running code.

Advantages:
• Ensures sponsor programs work.
• Allows sponsors to program complex sponsor-specific endpoints or derivations and review the outputs on their schedules.

- Allows for an anticipated format of real randomization and other unblinding information.
- Higher level of QC possible as testing programs are also available.

Disadvantages:
- Requires time for SDAC staff to be trained on the system and procedures.
- Requires quick sponsor IT responses to ensure appropriate staff have appropriate access.
- Limited ability and understanding of SDAC on programs – creating challenges if ad hoc outputs required.

3. SDAC receives Sponsor code that creates TLFs and simply merges on real randomization in own environment and runs code.

Advantages:
- Allows sponsors to program complex sponsor-specific endpoints or derivations and review the outputs on their schedules.

Disadvantages:
- System differences can contribute to difficulties in executing programs.
- Concatenation programs (combining individual outputs into a single easily-reviewed document) are often system dependent – and extra time should be allowed to ensure a smooth concatenation process can be achieved.
- If programming intended for CSR or safety reporting is re-used, it may not thoughtfully handle unclean data.
- Requires precise specifications about how to incorporate real randomization and any other unblinding information – the presence of this data may require programming modifications.
- Limited ability and understanding of SDAC on programs – creating challenges if ad hoc outputs required.

4. SDAC receives one of these options:

   (a) Sponsor's ADaM (industry-standard "Analysis Dataset Model") datasets and merges on randomization code and programs TLFs.
   (b) Sponsor's SDTM (industry-standard "Standard Data Tabulation Model") datasets and creates analysis datasets that are merged on randomization and creates TLFs.
   (c) "raw" datasets and creates analysis datasets that are merged on randomization and creates TLFs.

Advantages:
- Requires least amount of effort from sponsor.
- Does not require sponsor to mobilize efforts to support ad hoc requests.
- SDAC can silently facilitate ad hoc requests when knowledge of the requests could be informative to the sponsor.

- Allows SDAC programmers who are familiar with unclean data to specify algorithms specifically designed to accommodate unclean data.
- SDAC provides easy-to-read tables with key information programmatically highlighted to facilitate review of data.

Disadvantages:
- The most expensive option from the perspective of the sponsor.

There is an increasing level of ability of SDAC to reply intelligently and confidentially to DMC ad hoc requests in these different models. But there is also an increasing level of cost for SDAC services in these different models.

Hybrid models may be employed. For example, SDAC may do routine programming, and the sponsor may program complex efficacy outputs utilizing sponsor-specific analysis conventions for a formal interim analysis.

Thoughts should be made in advance in all of these approaches about how ad hoc requests from the DMC are handled, and in particular, how confidential ad hoc requests are handled. There must be a way that the SDAC can competently and discreetly create the needed outputs for the DMC. There are challenges for that taking place when the SDAC is using sponsor-provided code. The SDAC team will need to have a basic understanding of the code and data provided to them so that the DMC requests can be accommodated.

What Is Included in DMC Outputs?

David Kerr and Nand Kishore Rawat

Abstract This chapter is a general overview of these DMC outputs, clearly emphasizing that the layout and table of contents for outputs created for a DMC are expected to be very different from what would be generated at the end of the study for regulatory review. An overview of the safety data (e.g., adverse events, laboratory data), efficacy data (primary endpoint, or at least a proxy for it), and other data (e.g., demographics, disposition, exposure) is given. Proposal is also given for what is presented as tables, what as listings, and what as figures. Column structure and column naming are important, especially for studies with more than two treatments or more than one primary treatment phase. Suggestions are provided on the look-and-feel of the DMC package sent to the DMC so that it is both comprehensive, but also comprehensible.

Keywords Tables · Listings · Figures · Safety · Efficacy · Semi-blinding · Populations · Column structure · Categorical summary · Continuous summary · Program-wide DMC

The Data Monitoring Committee's (DMC) core responsibility of forming a recommendation is primarily based on the results contained in the Closed Report. This is a set of outputs generated, typically by treatment arm. The Closed Report should be a focused set of tables, listings, and figures to allow the DMC to interpret the essence of the study results with 1–4 h of review. It should be comprehensive, but also comprehensible. It should focus on key results, but hopefully not missing any unanticipated safety signal. The DMC can always request additional outputs if further details in a particular domain are required. Preferably, the Closed Report is no more than 200 pages. Individual outputs that are over 20 pages long should either be moved to an appendix or filtered or reformatted in some way to focus on key aspects.

D. Kerr
Seattle, WA, USA
e-mail: david.kerr@cytel.com

N. K. Rawat (✉)
Lantheus, King of Prussia, PA, USA

© The Author(s), under exclusive license to Springer Nature
Switzerland AG 2023
N. K. Rawat, D. Kerr (eds.), *Data Monitoring Committees (DMCs)*,
https://doi.org/10.1007/978-3-031-28760-2_9

37

The Closed Report should include high-level summary tables, a suitable use of graphics, and a minimum of listings. These outputs will focus on enrollment, demographics, baseline disease characteristics, exposure, disposition, safety (adverse events, deaths, laboratory data, vital signs, ECGs, etc.), and typically will also include efficacy data – either formal evaluation or informal evaluation to fully assess risk/benefit. A typical DMC report might include 20 tables – roughly one-third from demographics/disposition/exposure, one-third from adverse events, one-third other safety tables (labs, vital signs, etc.), plus one or two tables representing efficacy data. It might include figures – perhaps quite a few if lab data is shown over time in a helpful way, for example. Graphics of AEs can be very helpful as well. The DMC Closed Report typically would only include a handful of listings. Listings generally are not very helpful to the DMC. The DMC members can ask the SDAC for information on specific patients if needed. A reasonable proposal is that the DMC Closed Reports only have listings for serious adverse events (SAEs), Deaths, and Grade 3 or higher labs. There can be a push where the study team advocates for excessive number of outputs (AE tables filtered by a dozen factors, or a dozen listings) with the concern that the DMC must have everything – but that can have a negative impact by distracting (less experienced) DMC members from the more critical outputs and failing to recognize that the SDAC can produce additional outputs if requested by the DMC.

Specifics on outputs in the Closed Report are documented in subsequent chapters. Outputs should have clear titles, including the population. There should be a small number of distinct populations to avoid confusion – most commonly one that includes all subjects randomized (typically used for disposition and formal efficacy), and one that includes all subjects who have been treated (typically used for exposure and safety outputs). The output numbers should be simple (e.g. Table 1, Table 2, …), rather than what might be seen in an FDA submission (e.g. Table 14.1.3.1.1, Table 14.1.3.1.2, …).

Variables summarized in tables will typically follow one of three approaches:

- Continuous summary.

  - Mean, Median, 25th and 75th percentile, standard deviation, minimum, maximum.

- Categorical summary.

  - Make clear if subject can only be included in one line, or if subject can be represented in more than one line.
  - Many continuous summaries might also be represented as categorical. For example, age might be displayed using a continuous summary, but also displayed as <18, 18–65, and >65. This is especially true if there are certain continuous values that would trigger discussion of the DMC. The DMC and SDAC should collaborate on whether continuous data is most helpfully represented as continuous, categorical, or both.
  - Denominators are important to clearly explain when percentages are presented. Denominators may properly be a subset of the table population.

Consider a categorical summary of laboratory data at Week 12 (e.g. the count and percentage of those with Grade 3 or higher neutropenia at Week 12). Should the denominator be all subjects treated? All those with data at the Week 12 assessment? All those that were enrolled at least 12 weeks before the clinical cut-off date? All those that were enrolled at least 12 weeks before the snapshot date? Should the categorization be different for subjects who do not have Week 12 data for subjects who have not yet reached Week 12 vs. have reached Week 12 or withdrawn from study prior to Week 12? Answering this requires thought on the underlying question to be answered by the percentages produced. The study SAP could be useful place to look for answers, but that document likely would not fully address the real-time data that is part of DMC reporting.

- Subject-level category.

  - Single-line summary representing if a subject experienced the event of interest at least once.
  - Most typically for AE tables – each line will represent the number of subjects who had the specified AE at least once.

The Closed Report will typically show results split by treatment arm for randomized studies, or by dose group for dose escalation studies. The most common approach is to explicitly display the treatment or dose in the table columns and elsewhere (e.g., "Active" vs. "Placebo," or "20 mg" vs. "40 mg" vs. "60 mg"). Some sponsors (and a few DMC members) advocate to have outputs be semi-blinded. In that situation, the outputs would be labeled "Purple" vs. "Gold," or "A" vs. "B" vs. "C." In that situation, the DMC members could request to receive the explicit semi-unblinding decodes at any time (even at the time of the first meeting, although more commonly would be later in the study). The success of this approach depends on the DMC actively requesting the explicit semi-unblinding decodes if the recommendation that would be made would change depending on which arm was which. There have been examples where deaths were 5 vs. 15 (or similar), and the DMC assumed the smaller number of deaths were on the active arm and an encouraging signal, but rather the larger number of deaths were on the active arm and potentially action needed by the DMC. The semi-blinding approach is particularly ill-suited for studies with more than two arms, such as dose ranging studies (e.g. "Placebo" vs. "20 mg" vs. "40 mg") where the DMC should monitor for dose effect. In general, it is preferable to simply and explicitly give the treatment codes. Or, perhaps, in a two-arm study for security's sake have the labels be semi-blinded, but at each meeting (either provided separately in advance, or at the beginning of the Closed Session) provide the explicit semi-unblinding decodes. Do not switch semi-blinding treatment coding from meeting-to-meeting, which would impact the DMC ability to track imbalances over time. Semi-blinding is particularly challenging when randomization is not done in equal allocation (e.g., when randomization is 2:1 or 3:2:2:2). In these situations, semi-blinding would not be effective in hiding the some or all of the treatment arm information. Additional blinding of outputs can be attempted in these cases (e.g., remove all subject counts and only include percentages for categorical summaries), but the effort involved is not worth the limited benefit.

Many studies are simple where a subject continues on a specific treatment until completion of the study. However, many studies are more complex. There might be a lead-in/induction phase, maintenance phase, cross-over, or an open-label extension. How best to represent output from different phases of a study to the DMC is a common discussion during the Kick-Off Meeting. For example, a study that has a 12-week double-blind phase (active vs. placebo), followed by a 40-week open-label phase (all subjects on active), might be represented by having outputs in two parts, e.g. (Tables 1 and 2):

The DMC might conceivably want a Table 3 also, which includes all AEs combined into one table, split by original randomized treatment. (One final approach for Table 4 would summarize the full placebo experience – the first 12 weeks for those randomized to placebo – vs. full active experience – all experience for those randomized to active and the experience from Week 12 onward for those randomized to placebo. But this table would need to adjust for exposure time for any helpful comparisons.) The exact layout could be dependent on whether AEs are more or less of interest that occur shortly after treatment is initiated, or whether long-term AEs are more of a concern. Modern studies can be even more complex, with subjects who are responders (or non-responders) being re-randomized or with cross-over to

Table 1 Summary of AEs in double-blind phase (AEs through Week 12)

Active
Placebo
Total

Table 2 Summary of AEs in open-label extension (AEs from Week 12 to Week 52)

Active → Active
Placebo → Active
Total

Table 3 Summary of AEs in study (through Week 52)

Randomized to Active
Randomized to Placebo
Total

Table 4 Summary of AEs in study

Active (AEs through Week 52 for those randomized to Active, and AEs from Week 12 to Week 52 for those randomized to Placebo)
Placebo (AEs through Week 12 for those randomized to Placebo)
Total

specified open-label treatment at certain parts of the study. In all cases, the DMC should be aware of the Intent-to-Treat (ITT) principle to summarize patients based on how they were randomized, and understand the column structure they are presented, and whether it is a randomized comparison or not. (Even if there is ad hoc cross-over, typically the DMC will continue to review results by randomized treatment to maintain the ITT comparison.) Keep in mind that the sponsor might take ownership of monitoring open-label extension studies/phases. And the DMC duration might conclude at the time the randomization portion is locked and unblinded, even if the open label extension phase is ongoing.

A focused Closed Report is particularly important if the DMC is reviewing multiple protocols at one meeting (a "program-wide" DMC). Consideration needs to be made to have consistent naming/numbering/format of the outputs so that the DMC can digest results of two to eight protocols without requiring 2–8 times the effort. The value of the program-wide DMC is that the DMC deeply understands the new treatment, can give globally consistent requests, and can obtain earlier recognition of global trends of potential concerns. There are efficiencies, although a larger DMC will typically be required to cover all of the clinical disease areas of the studies – and more time will be needed by the SDAC to prepare and the DMC to review all of the outputs. Longer meetings likely will be needed, and efforts made to ensure there is no confusion within the DMC about which study is being discussed during the meeting. Review of each specific study might be rushed – one option is to only review a subset of studies at each meeting, perhaps in combination with having more frequent meetings. On occasion a "meta-analysis" is created in a program-wide DMC, but that has focused on just a few tables (e.g., the AE overview table). Many program-wide DMCs include studies which have different comparisons and maturity and length and patient population, and so simply combining all data together can be deceptive.

DMC outputs are from a snapshot in time. Both the sponsor and DMC should be aware of the implications of using real-time data. Chapter "What Data Is Used for DMC Outputs and Who Programs?" discussed the balance of currentness vs. cleanliness of the data snapshot. DMC outputs should summarize data to distinguish missing data for what is likely to never be available vs. "not yet available." For example, there might be 50 subjects randomized, but only documented dosing from 45 subjects. It is helpful for the DMC to know which of those five subjects were randomized shortly before the snapshot (and therefore likely were dosed, but simply did not have data submitted yet) and which of those five were randomized quite some time in the past (and therefore seem likely they truly were never dosed).

Real-time data will have some visits entirely missing (some that might be obtained in the future, and some that will never be obtained) and some missing data within visits. There may be clearly incorrect data (commonly from bad units – e.g., 1.83 cm height, rather than 183 cm height). There may be inconsistencies in the stratification data between randomization data and the Case Report Form (CRF) data. There may be inconsistencies within a patient's data (e.g., a patient has data on the AE form indicating that an AE led to discontinuation of treatment, but on the disposition form, there is no indication that the subject has discontinued treatment).

There may be AEs that have not yet been coded to a standard dictionary of medical terms, and there may be unadjudicated events. Most of these issues will not impact the DMC's ability to make an informed recommendation. There should not be over-reaction from the DMC or sponsor on these issues. Clearly, the DMC should expect a certain level of cleaning and completeness. And the SDAC might emphasize that the DMC review a median value, rather than a mean value, in the presence of an outlier clearly due to a bad unit. But the core focus of the DMC on AEs and other key safety data can still be done in the presence of real-time snapshot of data.

The analysis from the SDAC might be different from what would be done for final analyses by the sponsor, without unduly impacting the DMC's ability. For example, the final analysis might summarize outputs by "treatment received" rather than "treatment assigned," but for DMC purposes, there is little impact using the simpler "treatment assigned" with the assumption that >99% of subjects are only treated with the treatment assigned, rather than taking an incorrect treatment at some point. And the sponsor might use advanced data imputations or visit window-ing that would not affect >99% of dates or visits, and again DMC outputs can proceed with a simpler approach without any substantive impact. The above discussion does not necessarily apply to formal evaluations of efficacy, where there is typically the expectation of completely cleaned data and using the exact statistical algorithms that would be place at the time of the final analysis.

What Do the Final DMC Outputs Look Like and How Is It Delivered?

David Kerr and Nand Kishore Rawat

Abstract This chapter continues along those lines for how final materials are packaged and distributed to the DMC. Additional thoughts on the look-and-feel of the DMC package are provided, so that the DMC can easily but securely receive and – once received – easily review materials. An Executive Summary might be part of this DMC package. Pros and cons about going beyond static outputs are discussed.

Keywords Report delivery · Report package · Executive Summary

The Data Monitoring Committee (DMC) outputs are distributed electronically. Most typically outputs should be a single .PDF with bookmarks and hyperlinked table of contents. There have been examples seen of DMC members receiving 50–100 individual .LST or .RTF files – made even worse with cryptic names like "T01" and "T02." That is unacceptable. The DMC must be able to quickly review and annotate the results they are provided. Less egregious, but still unhelpful are files or bookmarks that say "t-demog" or "l-sae" rather than explicitly being named. In the past, paper binders were created in three-ring binders and delivered in advance of the meeting. Those days are now gone except on rare occasions. The vast majority of DMC members are happy to get reports instantaneously and securely, without worrying about delays in a mail room or a missed courier delivery at home. Electronic access allows for updated materials and minutes and other secure documents to be sent, and for easy searching for particular phrases within the documents. Statistical Data Analysis Centers (SDACs) should be able to support a secure file transport portal so that the members have unique accounts and passwords to allow access to the materials when needed. Reports typically are distributed 1 week prior

D. Kerr
Seattle, WA, USA
e-mail: david.kerr@cytel.com

N. K. Rawat (✉)
Lantheus, King of Prussia, PA, USA

N. K. Rawat, D. Kerr (eds.), *Data Monitoring Committees (DMCs)*,
https://doi.org/10.1007/978-3-031-28760-2_10

to the DMC meeting, although that can be updated to give more time (e.g., for a program-wide review with many studies to review) or less time (e.g., enrollment on hold until DMC recommendation, or formal interim analysis with need for a very quick response).

There can be concern of security. This was true even for the three-ring binders provided in past years (what if lost in mail room, or taken from front step of home, or left on an airplane, or carelessly left out in an office). But the issue is still valid when sent electronically. The assumption is that DMC members take their responsibility of confidentiality of materials seriously and treat these materials like they would HIPAA-protected information with appropriate electronic security and carefully making sure that no further electronic distribution is made. Some sponsors may insist on additional safeguards, although these come at the expense of convenience for the DMC. One approach is to prohibit downloads of the DMC outputs – the reports can only be viewed on the secure portal but not downloaded. DMC members have resisted this restriction, however. Another approach is to insist that DMC members delete DMC outputs from their computers within 2 weeks of the DMC meeting (after the minutes have been drafted, reviewed, and finalized).

The SDAC typically will also provide the most recent copy of the protocol and DMC Charter for easy reference. The meeting minutes from the previous meeting should also be distributed. Other study documents might be distributed. For example the Investigator's Brochure (IB) could be useful for the DMC to reference for rates of side effects seen in previous studies. The Informed Consent Form might be checked to see if an observed risk seen in the study is already described to incoming participants. The study's Statistical Analysis Plan might also be useful as well for the DMC.

The SDAC may produce an executive summary. The executive summary should not editorialize. The executive summary, if created, will summarize items of note from previous discussions and provide the current results on those topics. Newly created outputs or substantively updated outputs will be mentioned. The SDAC might independently flag imbalances or other results of interest. It could include an overview of the protocol synopsis and the definition of the populations analyzed. Additional details on the snapshot date, or the precise definition of non-trivial variables (e.g., "treatment emergency adverse events"), are also commonly seen. If an interim analysis, the specific details of the statistical criteria will be provided. The review of the executive summary by the DMC in no way absolves the DMC members from doing their own thorough review of the outputs. The author of the executive summary likely lacks the clinical expertise of the membership – there may be a 0 vs. 4 imbalance hiding in the outputs that warrants extensive discussion that was entirely missed in the executive summary.

There have been discussions for the past decade about going beyond static outputs. This could be providing the analysis datasets to the members, or applications that allow for real-time graphical investigation by individual DMC members. This is potentially helpful for the DMC, but many concerns remain that need to be addressed. Sending the datasets could be a concern because of worries of handling and distributing patient data outside the sponsor and SDAC. Even though no or minimal personally identifiable information is included, there still may be issues with datasets being sent to the DMC. Another concern is that these investigations likely would focus on the individual level analysis, rather than the group level analysis. There is occasionally the need to focus deeply on specific patients, but the DMC's key role typically is to look at by-arm results, not to micro-manage or adjudicate or second-guess the site's management of a subject. Another concern is that this would lead further into the issue of multiple comparisons. Even if the DMC could look at results by-group through an application, it would be easy to envision a DMC member looking at events by many different subgroups and become overly concerned with one subgroup that shows an impact, forgetting how many different approaches were first looked at. And finally, for documentation purposes, it currently is best that a single report can be provided to regulatory agencies at the conclusion of the DMC service documenting exactly what was reviewed to form the DMC decision, and the minutes will discuss exactly what materials all DMC members had access to. That would be challenging to document if there is a universe of different analyses each DMC member could do in advance of the meeting. Nonetheless, a few DMCs have been conducted that had access to web applications that allowed for interactivity, such as clicking a summary number on a table or a point on a scatterplot to get more information about the specific patients represented, or being able to create summaries (tables, listing or figures) based on user-defined filters. It is still unclear whether the use of these interactive web applications will expand and become the norm in the future.

What Types of Safety Outputs Does the DMC Receive?

David Kerr and Nand Kishore Rawat

Abstract This chapter goes into detail about the outputs that are both comprehensive and comprehensible for the DMC – this chapter focuses on safety. Details of the standard safety outputs are provided. These primarily are based on adverse event data, and a background is provided of how this adverse event data is captured and categorized. Detailed list of adverse outputs of most use to the DMC is given, with numerous examples of figures. Proposals for useful summarizations (including figures) from deaths, laboratory data, vital signs data, and other data sources are also provided.

Keywords Safety data · Adverse events · Serious adverse events · Adverse events of special interest · CTCAE · MedDRA SOC/PT · Deaths · Laboratory data · Liver function tests (LFTs) · Hy's Law · Vital signs

All post-baseline information provided to the Data Monitoring Committee (DMC) by treatment could theoretically reflect safety. Efficacy outputs will be discussed later, and in theory, the efficacy results could reveal safety concerns in the form of "reverse efficacy" – where the endpoint is showing a harmful unexpected trend. And disposition outputs could also reveal safety concerns (e.g., discontinuation from treatment or need for more concomitant medication could indicate harmful trend). This section will focus on the more traditional measures of safety, however.

The most common way to evaluate safety is through outputs of adverse events (AEs). Most commonly AEs are of interest when they are treatment-emergent – occurring

---

All figures created by Bill Coar.

D. Kerr
Seattle, WA, USA
e-mail: david.kerr@cytel.com

N. K. Rawat (✉)
Lantheus Medical Imaging (United States), King of Prussia, PA, USA

© The Author(s), under exclusive license to Springer Nature
Switzerland AG 2023
N. K. Rawat, D. Kerr (eds.), *Data Monitoring Committees (DMCs)*,
https://doi.org/10.1007/978-3-031-28760-2_11

after the first dose or intervention and including only subjects who have had at least one dose or had the intervention. On occasion, the DMC may want to look at a listing of AEs that were pre-treatment – for example, if part of the screening prior to starting treatment is to wean off a medication or an invasive scan is required, then the DMC might want to be aware of AEs during screening as well as the treatment-emergent AEs. The AE monitoring period typically extends through a set period of time after last dose (e.g., 28 days after last dose – or some number of days that represents four half-lives of the medication). The protocol and the sponsor and SDAC should clearly communicate what period of time is covered by the AE surveillance. Studies that have multiple parts (double-blind followed by open-label extension) should especially be clear on which AEs are summarized in which set of outputs. In some studies, particularly open-label studies, there might be a different schedule of visits for subjects on different arms. The more opportunities there are to ask a subject about AEs, the more likely to have instances of recall bias and therefore nominally higher rates.

AEs typically are categorized as serious (an SAE) or not. There is a formal definition of seriousness:

- results in death,
- is life-threatening,
- requires inpatient hospitalization or results in prolongation of existing hospitalization,
- results in persistent or significant disability/incapacity,
- is a congenital anomaly/birth defect,
- is a medically important event or reaction.

A core focus of the DMC will be on the by-arm summary of treatment-emergent SAEs.

AEs are typically categorized by severity grade – for example, mild vs. moderate vs. severe vs. life-threatening vs. fatal. Note that a severe AE is not necessarily serious, and vice versa. Grading might also be on a 1 vs. 2 vs. 3 vs. 4 vs. 5 scale. One standard grading approach is Common Terminology Criteria for Adverse Events (CTCAE). A standard output for the DMC is AEs summarized by maximum grade, and AEs that are Grade 3/Severe or worse.

AEs are commonly coded to MedDRA (Medical Dictionary for Regulatory Activities). It would be challenging to simply list all of the verbatim terms sites enter for each AE. Instead, a coding process is implemented to translate each verbatim term to a term in MedDRA. There are variants, but summaries of AEs are typically done with MedDRA at two levels – the System Organ Class (SOC) level which has 27 levels and then within SOC at the Preferred Term (PT) level which has nearly 20,000 unique terms. Note that due to the real-time nature of the data, some AEs might not yet be through coding at the time of the data snapshot. These should still be included in outputs, perhaps by showing the verbatim term entered. The DMC should be aware that very similar events might be coded into different PTs. It will not be immediately obvious if PT lines of a table can be added, or if simply summing the lines would lead to excess due to double-counting of subjects who show up on different lines due to having multiple events that were coded differently. For example, if two subjects show up in a table with PT of "Neutropenia" within the "Blood and lymphatic system disorders" SOC and two subjects show up in the same table with PT of "Neutrophil count decreased" within the "Investigations" SOC, it is impossible within just these results to determine if these two summaries represent two, three, or four unique subjects. The DMC can request the SDAC provide outputs that aggregate certain "constellations" of terms together. There are Standardized MedDRA Queries (SMQs) that do this for some standard groupings also. The study team may also have identified AEs of Special Interest (AEoSI). In such a case, the DMC outputs will include a summary of the PTs that are included in the set of AEoSI. One aspect that confuses DMC members who have not previously seen data summarized by MedDRA is how the condition under investigation is handled. They may suspect there is a problem when, in a Crohn's Disease study as an example, some subjects show up with a PT of "Crohn's Disease." The DMC member will say that it should either be 100% (because all subjects had the condition at baseline) or 0% (since it is not treatment-emergent). The answer is that these are recorded as AEs if the condition worsens, for example here, if there is a flare in the Crohn's Disease, it would be captured as an AE. In some protocols, a worsening of condition under investigation would not be captured as an AE but would be separately reported on the form that collects primary and secondary endpoints. The DMC should be informed on where and how AEs are collected as they relate to the clinical indication.

AEs can be categorized at the site for causality (e.g., possibly related, definitely related). It is suggested that the DMC generally ignore this categorization. The DMC will review the by-arm outputs. If there are more events on the active arm, then that type of event is likely causally related to the intervention. There is no need for the DMC to review or disagree with the investigator assessment of causality.

Information may be obtained if the AE resulted in interruption in treatment, change (reduction) in treatment, or permanent withdrawal from treatment. Note that some studies collect data if an AE led to discontinuation from the study – not just withdrawal from treatment. In most studies, this option should not exist. An AE certainly can motivate permanent withdrawal from treatment, but there's no reason that an AE should impact whether the patient stays on study and has data assessments collected in the future.

A standard set of outputs from AE data is shown here:

- Overall Summary of Treatment-Emergent Adverse Events.

  - At least one AE.
  - At least one SAE.
  - At least one Grade 3/Severe or higher AE.
  - AE leading to withdrawal from study drug.
  - AE leading to death.

- Treatment-Emergent Adverse Events by MedDRA SOC/PT.
- Treatment-Emergent Adverse Events by Descending Frequency of MedDRA PT.
- Treatment-Emergent Serious Adverse Events by MedDRA SOC/PT.

- Treatment-Emergent Grade 3/Severe or Higher Adverse Events by MedDRA SOC/PT.
- Treatment-Emergent Adverse Events Leading to Withdrawal from Study Drug by MedDRA SOC/PT.
- Treatment-Emergent Adverse Events Leading to Death by MedDRA SOC/PT.
- Treatment-Emergent Adverse Events of Special Interest by MedDRA SOC/PT.
- Treatment-Emergent Adverse Events by Maximum Grade by MedDRA SOC/PT.

These outputs are typically presented at a subject level – a subject will show up in the numerator if they have at least one of the events of interest – whether that be just a single event, or 2 or 5 to 10 events. Percentages represent the percent of subjects who had at least one of the events of interest. This typically isn't a concern, but a DMC might request additional information that provides insight into the total number of events on each arm, not just the number of subjects who had at least one event.

The DMC should be aware if there is differential premature discontinuation of treatment, the average on-treatment time could be different which again would impact the observed rate of AEs – but not due to any true difference in the AE profile. Another fine-tuning of AE outputs is to summarize AEs per 100 patient-years. This analysis will adjust if the average time under AE surveillance is different between the study arms. These analyses could include a subject at most once in the numerator or could include a subject multiple times if the subject had multiple of the event.

Typically, inferential statistics (e.g., p-values, confidence intervals) are not included in AE summary tables. Some DMC members have requested these, but there are concerns of misinterpretation. An AE summary table might go on for 10 pages, for example, representing all 200 unique preferred terms that have occurred at least once. Including p-values for each of these lines could easily be misinterpreted. Due to multiple comparison, one might expect – by chance alone – 10 lines to have p-value < 0.05. Looking at a p-value < 0.05 might be used as a flagging mechanism, but DMC members could easily mistake these AEs as demonstrating conclusive proof of difference.

The only listing of AE data typically included is the listing of SAEs. All other information can be provided on an as-needed basis by the SDAC. There is minimal value in extensive listings of AEs. One feature appreciated by DMCs is to include a cumulative listing of SAEs, but to highlight (in bold, or a different color font) the incremental SAEs that are new compared to the listing generated at the previous DMC meeting.

Figures based on AE data are less common but should become more common. Some examples are shown below. They can be helpful to look at differences, but do not entirely replace careful review of summary tables. An imbalance of 4 vs. 0 in progressive multifocal leukoencephalopathy (PML), for example, could be a critical topic for the DMC but likely would not stand out in the following graphics.

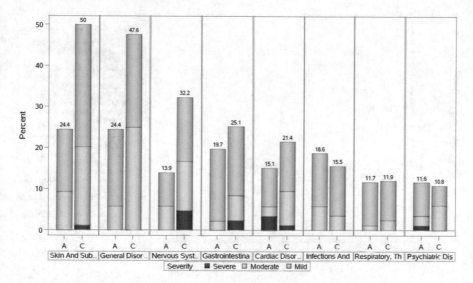

Fig. 1 Adverse events by System Organ Class

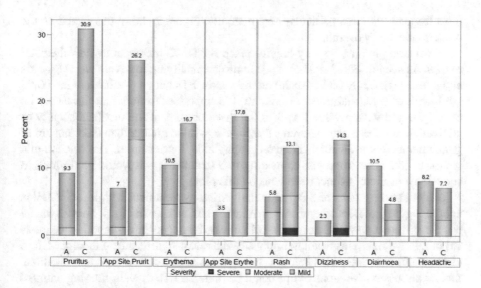

Fig. 2 Adverse events by Preferred Term

A plot of adverse event SOC in descending frequency, by treatment and indicating maximum severity, is a helpful way to quickly show results to the DMC as seen in Fig. 1.

This plot can similarly be done for PTs. This would take many pages, so filtering likely would be done. This could filter to only include the most frequent PTs overall,

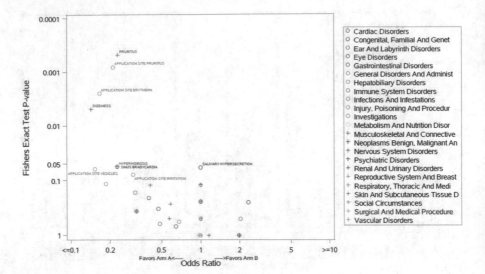

Fig. 3 Adverse events volcano plot

most frequent PTs in a particular treatment, etc. Figure 2 shows this sorted on the most frequent PTs overall.

A volcano plot can be very helpful place for DMC members to start their AE review. As seen in Fig. 3, it shows odds ratio (typically on log-scale) on the x-axis and p-value (typically on log-scale) on the y-axis. PTs with a p-value less than 0.05 are highlighted for additional discussion. It's important to note that there may be events flagged with p-value less than 0.05 that are not of interest (statistically or clinically), and there may be events that have a p-value greater than 0.05 that are of great interest (a 0 vs. 4 comparison on anaphylaxis, for example). The volcano plot is also of most use when just comparing two treatments – it would be difficult to show three or more distinct treatments on this plot.

A plot showing rates of SOC and relative risk by treatment within each SOC is another figure that DMC members gravitate toward, as seen in Fig. 4. Sorting can be done in different ways. The example below sorts by upper limit of confidence interval of relative risk. This is not to imply any statistical significance if the upper limit of the confidence interval is below 1, but again is acting as a filter to help facilitate further discussion. Versions of this could be done on PTs as well, filtering on most frequent or those with most difference (by HR, or by upper or lower confidence limit of HR).

Note that not all deaths will be AEs. For examples, deaths that are more than 28 days after the last dose might not be entered as an AE. And some studies have defined in the protocol that deaths due to disease progression are not entered as AEs. Deaths should be summarized for the DMC, but this information might be in two locations – one from the AE data, and one from a different data source of all deaths, or from the end of study or disposition data. The summary of death may show categorized reason for death, and may indicate which were within, say, 28 days of last

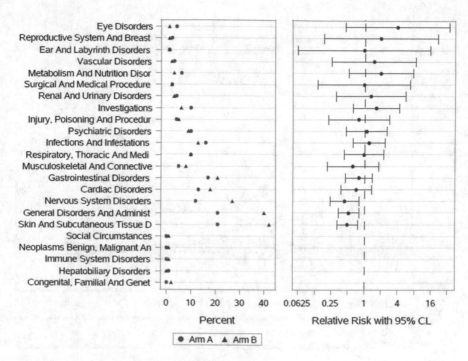

Fig. 4 Adverse events dot plot and relative risk

dose and which were beyond that time frame. A listing of deaths is commonly included as well.

Laboratory data is commonly provided, although this may not be as useful to the DMC as the adverse event data. The DMC may be more focused on laboratory abnormalities that have clinical consequence, in which case those will be captured in the AE data. Laboratory data is commonly categorized as either normal or as abnormal on a Grade 1–5 scale. Some lab parameters have a grading scale in two directions – one for the low ("hypo") values and one for the high ("hyper") values. The DMC may be interested in just one or both directions for these lab parameters. It is very easy and common mistake to have long, but unhelpful continuous summaries of lab data – repeating summaries for pages and pages for each time point within every lab parameter. The DMC must be provided more helpful outputs. If needed, the DMC can request additional materials from the SDAC.

A simple approach is to summarize maximum post-baseline grading for lab parameter (including "hypo" and "hyper" summaries separately). This is a short output and distills the most important features – looking to see if one arm or another has an excess of worst grades. A helpful figure for the lab data is a box-and-whisker plot which also includes means over time, as seen in Fig. 5. One figure per lab parameter is quick to review and shows visually both the overall trends (mean over time) as well as extreme values (points outside the whiskers). It is common to show a second figure per lab parameter representing the change from baseline. The

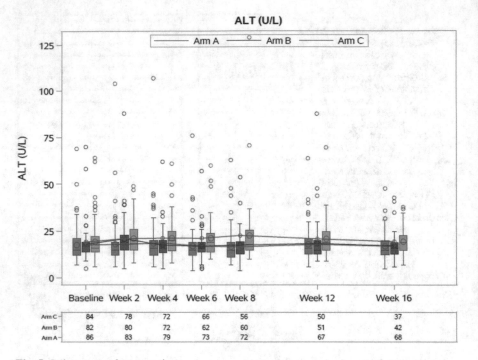

Fig. 5 Laboratory values over time

box-and-whisker plots likely will only include results from nominal visits, not any unscheduled visits. However, the table of maximum post-baseline grading will include all visits, including unscheduled visits.

If lab data is summarized in a table over time as continuous data, include change from baseline results and ensure outputs are in consistent units, using the units expected by the DMC members (which might be SI units, or might be U.S. conventional units). A summary over time might also include categories for values that would cause the DMC to have additional discussion.

Liver function tests (LFTs) – including ALP (alkaline phosphatase), ALT (alanine transaminase), AST (aspartate transaminase), and bilirubin – are a particular concern of DMCs because many treatments are known or suspected to cause hepatotoxicity. It is very common to have a distinct table summarizing number of subjects who have at least one ALT≥3xULN, ≥5xULN, etc. eDISH (evaluation of drug-induced serious hepatotoxicity) plots are a convenient way to graphically assess ALT and AST vs. bilirubin values. Values are assessed standardized compared to multiples of upper limit of normal (ULN). Both the table and figure will help the DMC to assess if Hy's Law laboratory criteria have been met (ALT or AST ≥3xULN simultaneously with bilirubin ≥2xULN).

Figure 6 shows the maximum post-baseline AST, ALT, and ALP vs. maximum post-baseline bilirubin. Each subject in only included once. A value in the top-right quadrant for AST and ALT plots might meet Hy's Law laboratory criteria. However,

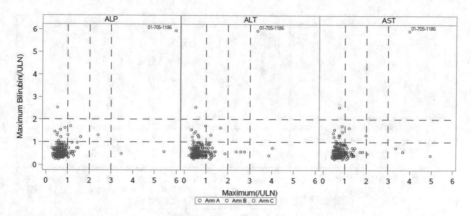

Fig. 6 eDISH plot

there is a chance that values in the top-right quadrant reflect elevations that were not synchronous.

A similar figure could be created that includes every visit. However, matching up ALT and AST visits vs. bilirubin visits to ensure they were synchronous is not always trivial if there are repeat assessments at a visit or unscheduled visits. It seems more common to present the maximum values of the parameters, and then investigate the specific patients of interest – those in the top-right quadrant – to see if elevations were synchronous.

If there are a small number of subjects with LFTs of interest (e.g., have met laboratory criteria for Hy's Law), a patient profile plot can be helpful, as seen in Fig. 7. These track multiple lab parameters over time (relative to multiples of upper limit of normal for each parameter). It can easily be seen if elevations are synchronous, and if elevations persist or are short-lived.

Looking at shifts from baseline to maximum in a table is helpful (perhaps looking at maximum toxicity grade vs. baseline toxicity grade). But a figure can also be instructive, as seen in Fig. 8. Here's an example looking at baseline vs. maximum value and highlighting subjects who have more than a 2xULN maximum. It's easy in this plot to see if these subjects are in the top-right corner of the plot which would indicate being abnormal at baseline, compared to the top-left corner of the plot which would indicate a new lab toxicity.

Listings of lab data can become too long very quickly. A helpful approach is to only list results that are Grade 3 or higher. Include other results from that lab parameter for that subject as well, so that the DMC can easily see the values that preceded and followed the high-grade lab result. Highlight (in bold or in a different color font) the high-grade value that triggered the patient's lab parameters being included in the listing.

Vital signs are not usually of interest unless the study is specifically intended or known to impact systolic blood pressure (SBP), diastolic blood pressure (DBP) or heart rate (HR). If of interest, include summaries that are similar as for lab data. Summarize the number of subjects who have had at least one value of certain

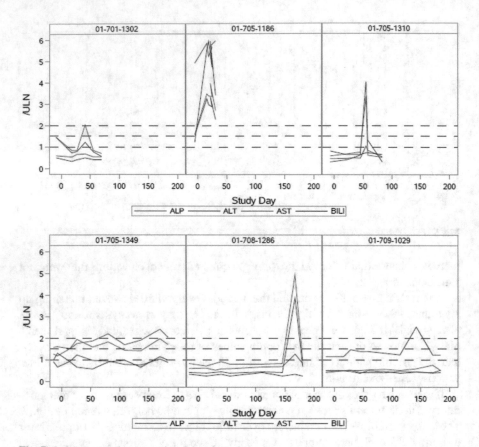

Fig. 7 Laboratory values over time by patient

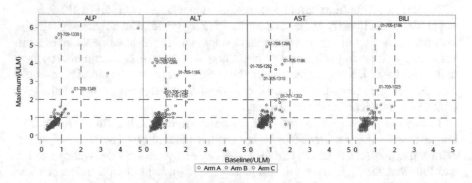

Fig. 8 Laboratory values baseline vs. maximum post-baseline

critical thresholds (e.g., SBP > 180 mmHg with an increase >20 mmHg from baseline) and include a box-and-whisker plot of SBP, DBP, and HR over time. Temperature and weight are typically not an informative way to address any safety concern, although those outputs may yield interesting results (e.g., increasing weight is a sign of efficacy in studies of patients with Crohn's Disease, and short-term summary of temperature might be of interest in a vaccine study). But in general, summaries from the AE outputs of terms such as "Weight decreased" or "Pyrexia" or similar would yield more informative safety results than from the vital signs dataset.

Other data might be included as needed for the study (e.g., QTc, ECG). The sponsor and DMC should always remember though that the DMC outputs do not need to include every piece of data collected. Generally, a summary of AEs (and SAEs in particular) will suffice instead of including tertiary safety parameters.

Kaplan–Meier figures are commonly used for presenting efficacy data where the endpoint is time-to-event data where some subjects have experienced the event and others are censored without yet having experienced the event. This is commonly seen for endpoints such as time to death, or time to disease progression (or death). However, Kaplan–Meier figures can also be used to represent safety data in helpful way for the DMC to reveal information about the time pattern of the events. For example, a Kaplan–Meier figure of time to first serious adverse event, as seen in Fig. 9, helps reveal if these SAEs are primarily early in treatment or evenly spread out over time and that could impact DMC recommendations on how to reduce the

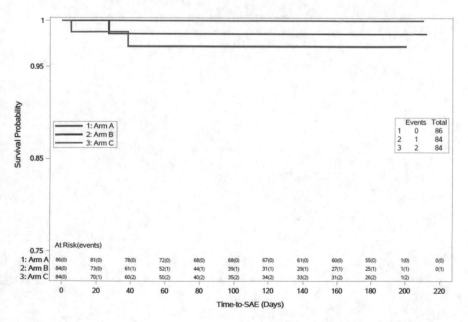

Fig. 9 Time to first Serious Adverse Events

risk of these in the future. The example below has only had three SAEs in the study, but might be more informative as the study matures.

What Types of Efficacy Outputs Does the DMC Receive?

David Kerr and Nand Kishore Rawat

Abstract This chapter goes into detail about the outputs that are both comprehensive and comprehensible for the DMC – this chapter focuses on efficacy. We advocate for efficacy data to be provided – even if non-inferential and just a proxy for the primary endpoint. This aids the DMC in a more comprehensive assessment of risk-benefit to provide more suitable DMC recommendations. The hazards of repeated assessment of efficacy data are noted, as are the hazards of overinterpreting early trends or lack of early trends. The principles of interpreting p-values are provided.

Keywords Inferential efficacy · Non-inferential efficacy · Risk-benefit · Endpoints · p-values · Kaplan-Meier figure · Repeated assessments · Alpha hit · Alpha spending

Non-inferential efficacy is a critical component to many Data Monitoring Committee (DMC) outputs even for meetings that some might ascribe as 'safety reviews'. Without access to some form of efficacy, the DMC cannot fully assess risk-benefit. The result would be the DMC recommending many studies stop for safety concerns when there is actually a good chance of efficacy eventually being demonstrated that more than offsets the safety concern. For example, many chemotherapy treatments are 'unsafe' given the severe side effects that are seen but are valued by the clinical community because of the efficacy provided.

This philosophy regarding the inclusion of efficacy data (even if non-inferential) is most compelling if the endpoint itself is clinically compelling – one that measures

---

Figure by Bill Coar.

D. Kerr
Seattle, WA, USA
e-mail: david.kerr@cytel.com

N. K. Rawat (✉)
Lantheus Medical Imaging (United States), King of Prussia, PA, USA

N. K. Rawat, D. Kerr (eds.), *Data Monitoring Committees (DMCs)*,
https://doi.org/10.1007/978-3-031-28760-2_12

how the subject 'feels, functions, and survives'. An endpoint that is a biomarker that has not yet been confirmed as a correlate for a clinically compelling endpoint is not as necessary to include. In these cases, the DMC might exclusively use the safety data to guide their recommendations.

The non-inferential efficacy can be a proxy to the primary endpoint. It could use data that is not fully cleaned or complete. There could be many approximations or other issues. But the purpose is simply to assuage DMC concerns that, in the presence of safety concerns, there is still some offsetting benefit or hope for benefit. As an example, in a study where the primary endpoint is disease progression confirmed by adjudication committee (e.g., BICR), it might be sufficient to have the DMC outputs include the site-proposed disease progression, given the expected high correlation between site-proposed disease progression and the disease progression confirmed by the adjudication committee. Or the primary endpoint might be a composite, but the DMC is presented with just the individual components of the composite. Or the endpoint is a time-to-event analysis, but the DMC is provided with the simple counts of the events, rather than being provided the Kaplan–Meier curve.

Including information on death can be viewed as both a safety output, as well as an efficacy output. It is typical to include an ad hoc Kaplan–Meier figure of OS. Note that the safety output will include a table using treated population, whereas the efficacy section may also include table of deaths using randomized population which could additionally include deaths from subjects that were never treated. It is relatively unlikely, but there could be concerns of 'reverse efficacy' where the trends in the endpoint variables are in the unexpectedly opposite direction and be indicative of safety concerns. That is another reason to include some measure of efficacy to the DMC.

There have been statistical concerns stated about including efficacy data as a routine matter. That is true for both non-inferential efficacy data but especially if inferential statstics are included (hazard ratio and confidence interval and p-value, for example) due to a perceived 'alpha hit'. There is a valid statistical concern that repeated assessments of data with the potential of early stopping do need to be accounted for in assessing p-values at interim and final analysis. But these efficacy outputs are not for potential stopping and therefore do not impact the interpretation of the final p-value at the final analysis. However, if objections persist, a simple approach is to build in that the alpha-spent at each review is some extremely small value, say, 0.0001. The end result on the final p-value is negligible. This does account for the unlikely but theoretical possibility that amazingly good results are seen during the study and the DMC may feel ethically compelled to recommend informing the sponsor and allowing subjects in the study and beyond to get access to this efficacious treatment earlier. Of course, as with all DMC recommendations, there can and should be discussion with the Sponsor Liaison and others (such as regulatory agencies) before any final decisions are made.

Efficacy data will, of course, be presented if the purpose of the DMC meeting is a formal assessment of specified criteria. That criteria could be for recommendation of stopping early for benefit, recommendation for stopping early for futility, or for other reasons. In many cases, just one or two outputs are included (one table, and

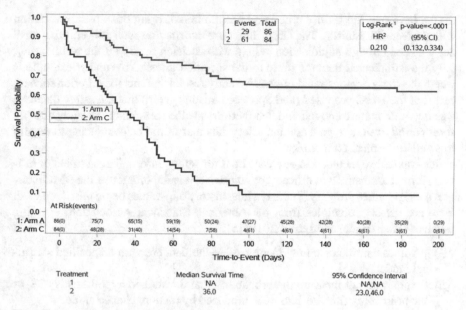

Fig. 1 Time to progression-free survival

one Kaplan–Meier figure, for example). Or it could be even more reduced – just to essentially one conditional power metric (likelihood of eventual success). The DMC will assess if the specified criteria were met and act accordingly, influenced (as agreed upon in the charter and previous discussions with the sponsor) by the totality of the data including the outputs focused on safety and study integrity. Fig. 1 shows what information might be contained in the Kaplan–Meier figure of time to progression-free survival (PFS) at a formal assessment of efficacy. Information in the top-right corner and the bottom might be removed if this PFS was being presented simply as a way to present efficacy as a possible counter-balance for safety concerns.

In other cases, additional efficacy data is provided to help the DMC fully assess the situation. This could include secondary endpoints and sensitivity analyses. The DMC, for example, would want information on the components of a composite endpoint to understand which of the components is driving a treatment effect (is it the most clinically compelling, or one that is less so). Sometimes, subgroups are presented in table or graphically as a forest plot so that the DMC can assess consistency of treatment effect across different subgroups. Some sponsors are dogmatic in how the DMC should operate – enforcing a binding recommendation on the data. Others are more flexible and provide the DMC the philosophy of the decision-making but encourage the DMC to use their full judgment. In either case, hopefully there is a Sponsor Liaison who can, if needed, have a confidential but frank discussion with the DMC if the recommendation to be made is not clearly obvious.

The DMC might see, for example, Kaplan–Meier curves where the curves cross and a trend for efficacy emerges later in the study follow-up. Based on that, the DMC might decide not to recommend stopping for futility even if numerically the results have met the futility criteria. Or, at minimum, bring this up for discussion with the Sponsor Liaison. The DMC might use information such as events that have not yet been through adjudication to help with decision making if the results using only the adjudicated data are close to the specified threshold. Another example is where the study demonstrated impressive early results that met the criterion for success, but the DMC proposed (and Sponsor Liaison agreed) that the safety database was not quite mature enough and that there was value continuing the study another short period of time to gain enough safety data that would be convincing to regulators and the clinical community.

If p-values are included in any way, be it for safety, informal assessment of efficacy, formal assessment of efficacy, or otherwise, it is important that the DMC interpret those p-values properly. There is a long history of p-values being misinterpreted. Here are the six principles from the American Statistical Association regarding p-values (Wasserstein and Lazar, 2016) [5]:

(i) P-values can indicate how incompatible the data are with a specified statistical model.
(ii) P-values do not measure the probability that the studied hypothesis is true, or the probability that the data were produced by random chance alone.
(iii) Scientific conclusions and business or policy decisions should not be based only on whether a p-value passes a specific threshold.
(iv) Proper inference requires full reporting and transparency.
(v) A p-value, or statistical significance, does not measure the size of an effect or the importance of a result.
(vi) By itself, a p-value does not provide a good measure of evidence regarding a model or hypothesis.

What Types of Other Outputs Does the DMC Receive?

David Kerr and Nand Kishore Rawat

Abstract This chapter goes into detail about the outputs that are both comprehensive and comprehensible for the DMC – this chapter focuses on outputs beyond safety and efficacy. These include outputs to evaluate study integrity, baseline data, disposition, and treatment exposure. Additional review that the DMC or SDAC might conduct includes event projections and randomization audits.

Keywords Study integrity · Enrollment · Study Disposition · Treatment Disposition · Demographics · Baseline disease characteristics · Protocol deviations · Exposure · Event projections · Randomization audits · Pharmacokinetic (PK) data

It is important that the study be well conducted and interpretable in a timely fashion. Other groups will help to oversee this, but the Data Monitoring Committee (DMC) has a role as well. And having a DMC meeting every 4 to 6 months, for example, is a good way that all teams involved in the study can periodically take a step back and consider if the study is on track, rather than discovering issues only toward the end of the study. Many of the following outputs are of most interest in an Open Session. There can be a healthy discussion in the Open Session if there is a realization that the population enrolled underrepresents a minority group, or there is an excess of protocol deviations, or a delay in obtaining adjudicated results.

Many of the following summaries are also of interest for by-arm results. For example, knowing that there are baseline imbalances in baseline disease severity could impact how the DMC assesses safety data in other portions of their Closed Session. And knowing that one group prematurely discontinued treatment more than another could indicate excess toxicity.

D. Kerr
Seattle, WA, USA
e-mail: david.kerr@cytel.com

N. K. Rawat (✉)
Lantheus Medical Imaging (United States), King of Prussia, PA, USA

N. K. Rawat, D. Kerr (eds.), *Data Monitoring Committees (DMCs)*,
https://doi.org/10.1007/978-3-031-28760-2_13

Items to include that are not specifically safety or efficacy related (although can provide insights into these domains) include:

- CONSORT diagram (see below).
- Information on participant screening – numbers screened.

  - Screen failure.

    - Reason for screen failure.

  - Still in screening.
  - Enrolled/randomized.

- Study accrual – in Open Session discuss if enrollment is ahead of schedule or behind schedule, and whether geographic distribution is as expected (not too many any one location or certain minimum needed at some locations – any caps or minimums required by geographic region, country or institution?)

  - by month (perhaps graphically over time – see below),
  - by location,

    - by global region,
    - by country,
    - by institution,

- Stratification (focusing on data used at time of randomization, but perhaps including Case Report Form (CRF) version as well).
- Demographics and Baseline disease characteristics – include the key factors that might have a prognostic impact on the endpoint.
- Key baseline laboratory values and other measurements.
- Previous treatment usage and other similar information.
- Eligibility violations (inclusion/exclusion criteria).
- Post-baseline major/important protocol deviations.
- Days between randomization and initiation of intervention.
- Adherence to medication schedule.

  - Duration on treatment.
  - Number off treatment (premature vs. completed) and reason why.
  - Number who have had at least one interruption, or reduction, etc.
  - Kaplan–Meier curve to treatment discontinuation can be useful.

- Participant intervention and study status.

  - Duration on study.
  - Number off study (premature vs. completed) and reason why (noting that very few people should be prematurely off study, other than for those who died, for well conducted long-term follow-up studies).
  - Kaplan–Meier curve to study discontinuation can be useful.

- Attendance at scheduled visits.

- Each expected visit – based on randomization date and cut-off date and the nominal timing of each visit.

  - Attended visit.
  - Did not attend visit, but should have attended visit.
  - Did not attend visit, since was previously discontinued from study.

- Currentness of data.

  - Time from last visit to data cut off for those still on study.
  - Time overdue from most recent visit attended to most recent expected visit for those still on study.

- Timeliness and completeness of adjudication of endpoints.

  - Time to endpoint being reported to site.
  - Time to endpoint being entered in database.
  - Time to endpoint package being sent to adjudication committee.
  - Time to adjudication committee making a final determination.

- Standard but not particularly useful – DMC members only rarely have comments on these tables when presented to them.

  - Baseline medications.
  - Post-baseline medications (unless very focused - for example post-baseline use of lipid lowering drugs in a cardiovascular study, or use of other anti-cancer regimens in an oncology study).
  - Baseline medical history.

- Options – key safety data by country – not necessarily to look at events by arm by country, but simply to see if the overall rates are similar or not. There have been examples of the SAE rate being very different between geographic regions. This may not introduce any bias necessarily in the by-arm results the DMC reviews in Closed Session, but it can be helpful for the DMC to know if overall rates are different across the regions.

A CONSORT diagram can be a quick way to present populations and disposition and exposure data to the DMC as seen in Fig. 1.

An enrollment figure over time can be helpful, as seen in Fig. 2.

Review of disposition can give useful information to the DMC – an excess of discontinuation from treatment on the active arm due to adverse events could indicate safety concerns for the active arm, or an excess of discontinuations due to disease progression on the control arm could indicate positive efficacy for the active arm. It is important to consider discontinuation from treatment as a separate variable from discontinuation from study follow-up. In most time-to event studies, for example, subjects should continue to have follow-up visits even in the case of premature discontinuation from treatment. The DMC should assess both by-arm differences as well as overall rates of discontinuation from treatment, and discontinuation from study follow-up. In open-label studies, the DMC will want to carefully review

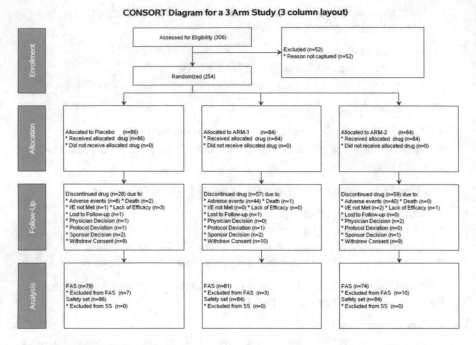

Fig. 1 CONSORT diagram

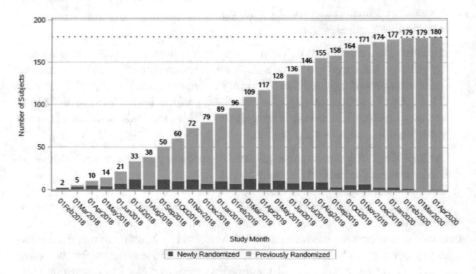

Fig. 2 Enrollment over time

early discontinuation from treatment and from study follow-up. Many studies have seen an immediate imbalance emerge with an excess of discontinuations on the control arm. In such studies it appears that the patients are enrolling, being randomized to the control arm, and then immediately withdrawing since they were not randomized to the 'desirable' active arm. This indicates that the Informed Consent process is not effective, and could impact the interpretability of the study. The options of the DMC to correct this are limited, but could include recommending additional training on the Informed Consent process to ensure that all prospective patients are truly amenable to participating on any of the randomized treatment options.

Duration on treatment is useful for the DMC to assess, as many studies only collect safety data through a certain period after last dose (e.g., non-serious adverse events are only collected through 28 days after last dose). Therefore, an imbalance of duration on treatment impacts the interpretation of safety data, in particular adverse event data. Alternative approaches to presenting safety data may be needed in the presence of imbalances in duration of treatment (e.g., AEs per 100 patient-years).

It is important for the DMC to review demographics and baseline disease characteristics. In smaller studies, there may be baseline imbalances that will affect the interpretation of safety data. One group may have subjects that are in a more severe categorization, or more subjects are from a particular geographic region. In the face of early baseline imbalances, the DMC should discuss in Closed Session whether are prognostic implications to these that would affect the later review of safety outputs.

The DMC is not usually charged with assessing the information fraction of the primary endpoint (e.g., how many events have been seen so far compared to the final number of events expected, and assessing when the number of events needed for interim analysis or final analysis will occur). That typically is done by the study team. The DMC can, and should, ask about timing and question the study team if it becomes clear the study will take appreciably longer than expected (either due to slow enrollment, or slow accrual of events in a time-to-event study, or for both reasons).

Occasionally, the DMC or SDAC will be asked to help provide input on the event projections. These are not necessarily based on confidential randomization information but might be based on other confidential information. An example would be a three-arm study – control vs. monotherapy vs. combination therapy. The study team might be most interested in the two pairwise comparisons of monotherapy vs. control and combination therapy vs. control. There might be 300 events required in each pairwise comparison of active therapy against placebo. If there were a total of 450 events, it could be that both comparisons have the needed number of events (e.g., 180 vs. 135 vs. 135, yielding 315 in each pairwise comparison), or neither comparison (e.g., 130 vs. 160 vs. 160, yielding 290 in each pairwise comparison), or one (e.g., 140 vs. 145 vs. 165, yielding 285 and 305 in the pairwise comparisons). Another example might be if a key biomarker is blinded to the study team. The study might want a certain number of events in the ITT group, but also a certain number of events from subjects who had baseline results that exceeded a key blinded biomarker. The DMC or SDAC can be provided the data (both endpoint and biomarker) and periodically give updates to the study team so that study activities can prepared in advance, without any knowledge of patient-level biomarker data.

There is inference that can be gleaned by the study team from knowledge of even this high-level information, especially if projections are provided on when the critical numbers of events will occur, so there should be a thoughtful pathway of communication of this information.

The SDAC may also assist the Sponsor with randomization audits, particularly early on in the study and particularly if there is a complex randomization schema. The SDAC will ensure that patients are randomized properly according to the master randomization schedule. The SDAC may check that the kits/assignments actually received match with the treatment assigned. The DMC is not necessarily involved or informed – although if a true process error is discovered during the randomization audit, then certainly the DMC should be made aware.

Some studies have provided pharmacokinetic (PK) data or had the DMC request PK data. This can be a challenge if PK data analysis is batched and only analyzed at the conclusion of the study or annually during the study. But it can be useful if the DMC is being asked to endorse dose escalation in early-stage studies. And some DMCs have requested PK data so that the SDAC can produce a summary of key AE outputs by PK level, to determine if the AEs might be caused by over-dosing and therefore help inform the DMC recommendations on what the best approach should be. Generally, however, DMCs do not receive or request PK data.

What About In-Between DMC Meetings?

David Kerr and Nand Kishore Rawat

Abstract This chapter mentions materials that might be sent to the DMC outside of the normal DMC meetings. These primarily are deaths, SUSARs, and/or SAEs. The pros and cons of these are weighed and what the process could be to help the DMC keep informed between DMC meetings without burdening them.

Keywords SUSARs · SAEs · Periodic safety events

Some data might be provided to the Data Monitoring Committee (DMC) in-between data review meetings. The most common of these is to pass along important safety events to the DMC. Examples would be providing listing or narratives of serious adverse events (SAEs) or SUSARs (Suspected Unexpected Serious Adverse Reactions) either in near real-time or on some periodic schedule such as monthly (in that case, perhaps cumulative or perhaps incremental – or perhaps both – since the previous transfer).

It is understandable that the DMC and the sponsor would want to keep the DMC apprised of important safety events that occur in the months between DMC meetings. However, the rationale is not always compelling. The DMC is at its most valuable when it is looking at by-arm results using cumulative data. There are typically very few types of events where a single one or two instances of the event would cause the DMC to have enough evidence to recommend action. (Examples of a single event causing consternation might be anaphylaxis, PML, or Hy's Law case – or a death in a reasonably healthy population.) Instead, it can lead to distraction from the DMC. A common response to a DMC receiving narratives from these events is to question the clinical approach taken by the site, or the site-proposed causality/relationship to study treatment. It is not the DMC's job, however, to manage clinical care of a specific patient or to concern themselves with site-proposed

D. Kerr
Seattle, WA, USA
e-mail: david.kerr@cytel.com

N. K. Rawat (✉)
Lantheus, King of Prussia, PA, USA

© The Author(s), under exclusive license to Springer Nature
Switzerland AG 2023
N. K. Rawat, D. Kerr (eds.), *Data Monitoring Committees (DMCs)*,
https://doi.org/10.1007/978-3-031-28760-2_14

causality/relationship to study treatment. The site, and the sponsor to a degree, takes on the responsibility for patient care. The DMC can remind sponsor and sites if protocol-specified procedures or region-specific best practices are not being followed – but that should not be based on single anecdotes.

Typically, these periodic safety events are provided to the DMC without treatment information. If requested, the Statistical Data Analysis Center (SDAC) should be able to provide the treatment information. (It is important that the SDAC has access to randomization information in real-time without needing to obviously request it. If a subject was recently randomized and the DMC felt it was critical to obtain the treatment, the SDAC should be able to get that randomization information without alerting the sponsor of that fact.) This would be important if, indeed, the safety event was important enough, unto itself, to possibly warrant a DMC recommendation for action. The DMC can call an ad hoc meeting if needed to review the case and propose action.

Less common, but likely more helpful, is to pass along a tabulation of safety data periodically. For example, a table of SAEs presented by-arm (using the proper denominators of all subjects at risk as of this time), accompanied by a listing of the incremental SAEs since the previous transfer could be helpful, especially early in the study and if the time between DMC meetings is lengthy.

The value of having periodic safety events provided is greatest early in the study, particularly if there is not yet much knowledge of how this new treatment impacts this patient population.

The process for these periodic safety events should be specified in the DMC Charter. These outputs might go to the full DMC, or just the DMC Chair, or just the DMC members who have the clinical expertise to review them (e.g., not the DMC statistician). There might be formal written acknowledgment from the DMC Chair that there was or was not action needed, although more frequently there is no such paperwork. The DMC Charter might encourage that the frequency or filter change as the study database matures. For example, the frequency might start monthly, but then go to bi-monthly after the first year. Or it might start by providing all SAEs, but after the first year only pass along fatal events.

There typically is no formal monitoring plan for these events. However, some studies have implemented plans. For example, the DMC might start reviewing Grade 3 infections starting after the fourth Grade 3 infection is seen, and be provided treatment arm information, with specific guidance for when an ad hoc meeting should be called based on by-arm difference (e.g., ≥ 4 vs. 0, ≥ 6 vs. 1, ≥ 8 vs. 2). The SDAC would assist in receiving information in real-time and unblinding cases and working with the DMC.

What Types of Formal Interim Analyses Does the DMC Review?

Lingyun Liu and Cyrus Mehta

Abstract This chapter is an in-depth discussion of formal interim analyses of adaptive clinical trials with examples, only gently touching on advanced statistical theory. Adaptive designs can include guidelines to stop the study or a treatment arm early, or to perform a sample size re-estimation. These adaptive designs can focus on stopping for overwhelming benefit, or for futility. In all cases, both the statistical methodology and the communication plan used must ensure integrity of study results. Numerous examples are provided to illustrate key concepts such as the O'Brien Fleming error spending function for generating early stopping boundaries, and use of conditional power, for futility stopping or to identify the promising zone for sample size re-assessment.

Keywords Adaptive designs · Efficacy · Futility · Sample size re-estimation · Adaptation committee · Type 1 error inflation · O'Brien-Fleming approach · Conditional power · p-value · Confidentiality of results · Promising zone · Information fraction · Decision boundaries

L. Liu (✉)
Lantheus Medical Imaging (United States), King of Prussia, PA, USA

Department of Biometrics, Vertex Pharmaceuticals, Boston, MA, USA
e-mail: lingyun_liu@vrtx.com

C. Mehta

Cytel Corporation, T. H. Chan School of Public Health, Boston, MA, USA

Harvard T. H. Chan School of Public Health, Boston, MA, USA
e-mail: cyrus_mehta@cytel.com

© The Author(s), under exclusive license to Springer Nature
Switzerland AG 2023
N. K. Rawat, D. Kerr (eds.), *Data Monitoring Committees (DMCs)*,
https://doi.org/10.1007/978-3-031-28760-2_15

Adaptive designs have been increasingly used in clinical trials as alternatives to traditional fixed sample designs over the past decade to improve the efficiency of clinical development. The FDA guidance on adaptive design in 2019 (FDA 2019) [5] defines adaptive design as a clinical trial design that allows for prospectively planned modifications to one or more aspects of the design based on accumulating data from subjects in the trial. The guidance document also highlights a few advantages for adaptive designs over non-adaptive designs including statistical efficiency, ethical considerations, improved understanding of drug effects, and acceptability to stakeholders. The efficiency of adaptive designs comes from the flexibility to modify the studies according to pre-specified rules based on cumulative data already observed from the trial itself. Adaptive designs can be applied to all phases of clinical development. The types of adaptations could include early stopping for efficacy or futility, sample size re-estimation, treatment selection, population enrichment, and response adaptive randomization. Sponsors or other parties involved in trial conduct are typically strictly blinded to the interim results for protecting the study integrity. To implement the adaptations in adaptive Phase 2 or Phase 3 confirmatory trials, an independent board is needed to review the unblinded interim data and make recommendations to sponsors. This role is beyond the traditional role for Data Monitoring Committee (DMC) which is to monitor the safety and protect trial subjects. FDA guidance (FDA 2019) [5], Sanchez-Kam (2014) [6], and Antonijevic et al. (2013) [7] have discussed two models for implementing the adaptation decisions. One model is to use another committee (adaptation committee) to review unblinded interim data and make adaptation decisions. The other model is to have one single DMC for both tasks: safety monitoring and adaptation decision. Although both models could work and there have been trials using both models, it has been recognized that the model with two separate committees could lead to conflict recommendations made by two groups. Therefore, a single DMC with all the needed expertise could be more efficient to monitor adaptive trials and make adaptation decision, and this is the more commonly used model in practice.

Regardless of the endpoint used for the study, formal interim analyses need to properly account for the repeated nature of conducting statistical tests prior to the final analysis. It is entirely incorrect to simply assess the endpoint at a nominal critical p-value of 0.05 at each meeting. Such an approach would inflate the Type 1 error – leading to studies to be recommended to stop early too frequently, simply by chance due to the repeated testing. For example, ignoring the correlation of repeated tests, if a study had two interim assessments and a final assessment – all done at the 0.05 level – the chance of a positive result would be 0.14 even in the presence of no treatment effect. On the flip side, one does not need to use critical values of 0.05/3 or similar algorithms that simply sum to 0.05 over the two interim and one final assessment. That would be overly conservative, resulting in substantial loss of power. The correlations between the p-values at the different meetings can be employed to find an efficient statistical approach that preserves that Type-1 error.

Various statistical approaches have been created to ensure that there is no Type 1 error inflation. The most common is the O'Brien-Fleming approach. The critical p-value at different interim analyses is adjusted based on the percent of information available and the amount of the Type-1 error that one is allowed to spend at that interim analysis time point. (This information fraction depends on the endpoint – for time-to-event studies it is the current number of events divided by the final expected number of events. For an endpoint such as Week 12 assessment, it will be the number of subjects who have that Week 12 assessment divided by the number of subjects to be enrolled.) The O'Brien-Fleming approach allows great flexibility in its input parameters to allow for more or less restrictive early stopping. But one does not get something for nothing. For a design with one or more interim analyses and an overall Type-1 error of 0.05, the final analysis will have a p-value less than 0.05 because some of the alpha will have been spent earlier. However, in many cases, it is a minimal amount, and the great value of early stopping is more than outweighed by having the final analysis based on a p-value criterion of 0.048 (for example). Error spending functions can be used for both assessing benefit and for assessing futility. One or both can be used at a formal interim analysis.

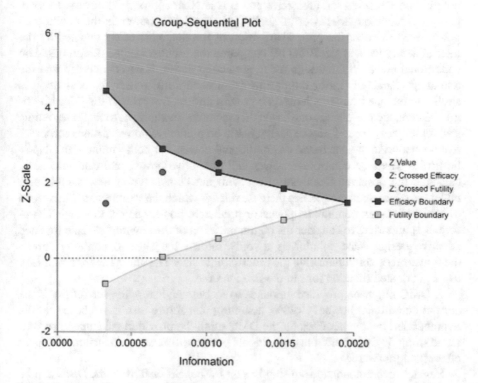

Fig. 1 Example of Interim Boundaries and Results Over Time

Figure 1 shows a typical funnel-like shape of the early stopping boundaries if both benefit and futility are assessed. These boundaries are generated by a popular error spending function called the Lan and DeMets O'Brien-Fleming error spending function. Due to its conservative nature, more dramatic results are required for early efficacy stopping at the beginning of the trial. In this example, at the third interim analysis, the results exceed the criteria to declare benefit so that the trial may be stopped early.

Conditional power is often provided to DMCs to assess the likelihood of success if trial continues. This is used most frequently when assessing futility. The conditional power asserts what is the likelihood of a successful result, given the current results. It is important to pre-specify and communicate in advance what the anticipated future results are based on:

- Assume hypothesized treatment effect going forward.
- Assume observed treatment effect going forward.
- Assume null treatment effect going forward.
- Assume an averaged treatment effect (between observed and hypothesized) going forward.

In practice, conditional power is often evaluated at the observed treatment effect. However, the observed treatment effect has variability for which conditional power cannot account for. Predictive power is another approach to govern the sample size adaptation (Mehta et al. 2022) [9]. Predictive power is the conditional power averaged over the prior distribution of the treatment effect updated by the interim data. Mehta et al. (2022) [9] compared the promising zone design based on conditional power to the design using predictive power. It was concluded that the operating characteristics of the design using conditional power for adaptation is similar to the one using predictive power with informative prior. If the prior is non-informative, the promising zone based on predictive power starts to increase sample size for larger z_1. Therefore, the design based on predictive power is more conservative compared to the one based on conditional power. For both sample size adaptation rules based on conditional power and predictive power, one can potentially calculate the observed treatment effect with the knowledge of new sample size under the continuous sample size increase rule for which the sample size is increased to achieve a target conditional power or predictive power subject to a cap. Therefore, it is important to consider the communication of the new sample size for such adaptive design. One solution is to only inform the sites to continue enrollment instead of communicating the exact sample size. The alternative solution is to use a flat or step function for sample size increase.

A DMC might be provided guidance to recommend stopping for futility if the current conditional power is <20%, assuming that future results are based on the hypothesized treatment effect. Or the DMC might be provided guidance to recommend stopping for benefit if the conditional power >90%, assuming a null treatment effect going forward.

Besides recommendations to simply stop for benefit or futility, the DMC can be used for other adaptations. The most frequently used are adaptions to the sample size. Other adaptations include dropping of treatment arms in a multi-arm trial and selection of a sub-group for going forward (also known as population enrichment). In what follows, we shall provide examples of these different types of adaptations.

Focusing on adaptations to sample size, algorithms can be provided as guidance to the DMC on increasing or decreasing the statistical information (quantified in terms of number of patients or number of events) needed to have a compelling answer to the question of interest – does this new treatment differ from a control arm? This is an appealing idea. One would hate to finish a study and have a p-value of 0.08 and wish that another 100 patients or events had been collected. There are statistical approaches that maintain statistical integrity when analyses along these lines are conducted by the DMC using access to unblinded results by treatment arm. Figure 2 shows graphically the philosophy of this, using the example from the Valor study (discussed further later in this chapter).

Recommendations from the DMC to the sponsor should generally exclude any details of the interim analysis results. In addition, the recommendation made by the DMC should be communicated in a way that does not convey, indirectly, information on the unblinded interim results. For example, in adaptive design trials with sample size adaptation, knowledge of the sample size adaptation algorithm and the new sample size could allow back-calculation of interim treatment effect size. This, in turn, might bias the study outcome if investigators having knowledge of the current estimate of treatment effect selectively exclude patients from the trial. For such designs, it is important that the communication on adaptation decision should be made in a way to minimize the information which might be inferred. Therefore, careful planning with respect to the information to be communicated after interim analysis is very important. There are alternative approaches to preventing the backward calculation/reverse engineering of treatment effect estimates by carefully designing the adaptation rules. For example, if the sample size is increased after interim analysis, trial sites could be informed that the targeted enrollment number has not been reached yet rather than being informed about the exact new sample

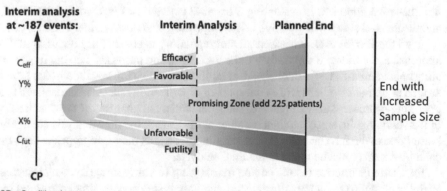

CP = Conditional power
The probability of success (statistical significance) at the end of the trial given current data trend

Fig. 2 Example of Promising Zone

size. Alternatively, one could use step functions for sample size increase rather than a continuous sample size increase rule to limit the knowledge that can be inferred from knowing the target sample size. Knowing that the study will go beyond its originally intended number of patients/events is informative, but hopefully minimally so. Even so, it would be best practice not to make public what exact statistical methodology for the sample size reassessment is being used.

Algorithms can be provided to the DMC to assess dosing. For example, a study might be initiated with placebo vs. low dose vs. high dose. Based on efficacy results halfway through the study, the DMC might be tasked with recommending

- Stop the study for futility (both active arms performing worse than placebo).
- Stop one active arm (the other active arm is performing appreciably better than both placebo and the other arm).
- Keep the study going (both active arms are doing better than placebo, but neither active arm is appreciably better than the other).

The DMC charter will need to state how a change in dosing or future enrollment is communicated – perhaps by restricting that information only to the vendor responsible for randomization and treatment and to the drug-supply vendor. It will need to be made clear how subjects enrolled on a stopped arm are handled – are they treated with their current treatment (with no further enrollment into that arm) or do they have treatment stopped entirely or do they cross-over to take the treatment that is continuing.

The DMC might be tasked to evaluate results and make recommendations on subgroup results – for example PDL1 levels in an oncology study. A recommendation might be made to change from enrolling "all comers" to instead only enrolling subjects with PDL1 > 5%. The subsequent statistical analyses will need to be carefully controlled for the fact that decisions were made midway through the study based on the results seen at this interim review.

The DMC might also be tasked to review results in a "seamless" Phase 2/3 study and evaluate whether to initiate the Phase 3 portion. A situation here might be where the Phase 2 component is evaluating 2 or more active doses against a control arm, and the Phase 3 study will continue just a single preferred dose against a control arm.

In all the above cases, it is essential that regulatory agencies fully understand the approach and endorse if the study will be used for filing purposes. Having the operating characteristics well established by simulation studies is a key factor. The DMC as well as non-statisticians within the sponsor organization must also understand the operating characteristics, and how to interpret statistical analyses at the conclusion of the study. Again, since nominal p-values at the time of the final analysis are likely biased due to the interim reviews and resulting decisions made, appropriate statistical adjustments must be made to the final analysis.

Exposure of interim results can be devastating to study integrity. Any updates to the protocol (e.g., entry criteria, statistical analysis) may be misinterpreted by investigators with access to this information, resulting in selective exclusion of patients. Sites may act differently – either slowing enrollment or enrolling different subjects or acting differently in treating patients or entering data after learning about the interim results. Therefore, DMCs should be very leery of sponsors who wish to conduct interim analyses for "business purposes," rather than to specifically generate a recommendation to stop for efficacy or futility. Some sponsors assert that they need to learn about trends at a certain time so that money can be raised for future studies, or to start building production facilities, or to start planning studies, or to trigger regulatory activities. DMCs should feel free to push back on this or, at minimum learn more about the communication plans/data access plans for how that confidential interim data will be handled. These plans should reinforce that all confidential information should be disclosed on a "need to know" basis only – and recognize the damage to study integrity that would result from that information being disseminated further within the company and outside the company.

Examples of Formal Interim Analyses Conducted by DMCs in Group Sequential and Adaptive Trials

Example: Group Sequential Design with a Single Primary Endpoint

- Study needs 400 progressions or deaths (note – number of actual subjects enrolled is irrelevant).
- Sponsor interested in assessing futility early, and both futility and benefit with data that is more mature.
- Endpoint is log-rank test of time to progression or death (censored for those still alive without progression), stratified, with hazard ratio < 1 indicating reduction in hazard in favor of experimental arm, overall alpha is 1-sided 0.025.
- Possible formal monitoring boundaries for DMC.

One possible set of stopping boundaries is shown in Table 1.

The statistical data analysis center (SDAC) may need to re-compute boundaries if the actual information is not exactly what was specified. For example, the protocol might assume the formal analysis takes place when 50% of information is reached (200 subjects had died or had disease progression, out of the 400 needed for the final analysis). However, at the time of the data snapshot for the interim analysis, there have been 210 subjects who have died or had disease progression. One approach would be to truncate analysis at the date at which the 200th subject had died or had disease progression. But most would agree that removing useful information is not helpful and would want to include all data. The boundaries would be recomputed to reflect the 53% of information currently available and that boundary used by the data monitoring committee (DMC) as it reviews information from all 210 events.

Table 1 Example of Stopping Boundaries with One Primary Endpoint

Look	Events	% Info	Futility if HR	Futility if 1-sided p-value	Benefit if HR	Benefit if 1-sided p-value
IA #1	200	50%	HR > 1.0	P > 0.50		
IA #2	300	75%	HR > 0.9	P > 0.30	HR < 0.7	P < 0.003
Final	400	100%			HR < 0.75	P < 0.024

Example: Group Sequential Design with Co-primary Endpoints

- co-primary endpoints of progression free survival (PFS) and overall survival (OS). Need 300 PFS, and 400 deaths. (More deaths are required because the expected treatment difference is smaller for evaluating deaths, hence more events are needed to have an adequately powered analysis),
- PFS co-primary endpoint will be observed sooner, as there are fewer events and they will accrue sooner by definition (since every OS is also part of PFS endpoint).
- Assess futility on PFS early, and then both futility and efficacy for PFS mid-study. OS will be evaluated for efficacy mid-study and at the time of final PFS analysis.

One possible set of stopping boundaries is shown in Table 2.

Example: Adaptive Design with Sample Size Re-estimation

- Promising zone employed to possibly increase sample size if results are not clearly going to show efficacy at the current sample size but is promising to be significant with an increased sample size.

  - Planned analysis with a total sample size of 442. Interim at 208.
  - If conditional power >0.8, leave at 442.
  - If conditional power 0.3–0.8, increase sample size to 884.
  - If conditional power 0.1–0.3, leave at 442.
  - If conditional power <0.1, recommend stopping the study for futility.

Four clinical trials will be presented next each with different types of design features. The REDUCE-IT trial is a real clinical trial which utilized group sequential design to assess efficacy at interim analyses. It illustrates the benefit-risk information DMC reviews at interim. The second example is VALOR trial which used adaptive design with sample size re-estimation to boost the study power. The ADVENT trial is an example of seamless Phase II/III design which combined Phase II dose selection and Phase III confirmatory testing in a single trial for registration purpose. The last example on TAPPAS trial has an adaptive enriched design which provides the option to enrich the study population based on interim results.

Example: Group Sequential Design-REDUCE-IT Trial:

Group sequential design is the most popular alternative design to fixed sample design in clinical research. Such design repeatedly assesses the cumulative data at interim analyses to allow early stopping for safety, efficacy, or futility. The criteria

Table 2 Example of Stopping Boundaries with Co-Primary Endpoints

Look	PFS Events	% Info	Futility if HR	Benefit if HR	OS Events	% Info	Futility if HR	Benefit if HR
IA #1	100	33%	HR > 1.0					
IA #2	200	66%	HR > 0 0.9	HR < 0.6	100	25%		HR < 0.5
Final PFS	300	100%		HR < 0.7	200	50%		HR < 0.6
Final OS					400	100%		HR < 0.75

for efficacy stopping must be pre-specified to control the overall type 1 error. The REDUCE-IT trial used group sequential design. The REDUCE-IT (Olshansky et al. 2021) [8] study was a randomized placebo-controlled cardiovascular outcomes trial in patients treated with statins, who had controlled low-density lipoprotein cholesterol, but persistently elevated triglycerides along with overt presence of or high risk for cardiovascular disease. The primary endpoint for this study was time from randomization to first occurrence of a composite of cardiovascular death, non-fatal myocardial infarction, non-fatal stroke, coronary revascularization, or unstable angina requiring hospitalization. The trial was planned to enroll 6990 patients and observe 1612 events for targeting 90% power to detect 15% relative risk reduction from an event rate by 4 years of 23.6% in the placebo group to 20.5% in the experimental treatment group assuming an 18-month enrollment period and a median follow-up of 4 years. To protect against the possibility that the actual placebo event rate is lower than estimated, an extra 1000 patients will be enrolled (approximately 7990 patients in total). One interim analysis was planned at 967 events which corresponds to 60% of the total number of primary endpoint events. The efficacy stopping boundary was determined based on O'Brien-Fleming error spending function. The critical boundary on the p-value scale at interim is 0.0076 and the final boundary is 0.0476.

The study enrolled 8179 patients in total and observed 1606 primary endpoint events. In addition to the first interim analysis at 60% events, a second interim analysis was taken when 80% events (1218 events) were observed with p-value boundary 0.0211. The final boundary is 0.0437 adjusted to account for interim analyses and the final observed total events of 1606.

The interim results of the study were reviewed by an independent DMC. The analyses were performed by the SDAC unblinded to the treatment assignment. A single DMC review board was planned to review both efficacy and safety data to make recommendation to the sponsor. The DMC included two physicians, a statisti-

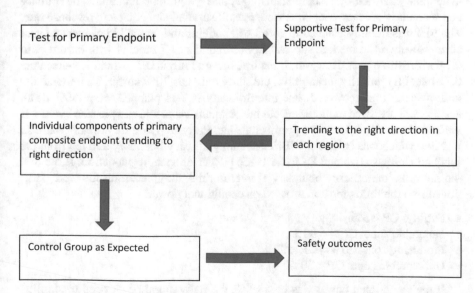

Fig. 3 Interim decision tree

cian, and a non-voting independent statistician. The DMC reviewed the interim data based on a pre-specified decision-making process including assessment of safety, treatment arm performance, primary endpoint analysis and internal robustness analysis. Figure 3 is the modified decision process for illustration purpose which were developed at the design stage to guide DMC for decision making. The DMC would review the primary test for the primary endpoint based on log-rank test. If the primary test is significant, the supportive test for the primary endpoint based on Cox regression model should be reviewed. If the supportive test is also significant, DMC should check the treatment effect in terms of hazard ratios in each region and make sure all the hazard ratios trend to the right direction. If treatment effects in all regions trend to the right direction, the next step is to check whether the individual components of the primary composite endpoint also trend to the right direction. DMC also need to check if the observed control group event rate was as expected by performing a meta-analysis with all studies involving the control treatment. Last, DMC needs to check whether there are any safety issues which warrant continuous follow-up of the patients. The DMC might recommend stopping the study for efficacy if all the assessment criteria in Fig. 3 are met. Although the study crossed the efficacy boundary at both interim analyses, DMC discussed the overwhelming efficacy results and considered historical examples of failed cardiovascular outcome studies for triglyceride lowering and mixed omega-3 therapies. The DMC recommended study continuation weighing in the importance of a more mature data set to support robustness of final efficacy and safety findings.

Example: Adaptive Sample Size Re-estimation-Valor Trial:

The Valor study is a phase 3, double-blind, placebo-controlled trial conducted at 101 international sites in 711 patients with acute myelogenous leukemia (AML). Patients were randomly assigned 1:1 to vosaroxin plus cytarabine or placebo plus cytarabine stratified by disease status, age, and geographic location. The primary and secondary efficacy endpoints were overall survival and complete response rate. The trial was initially planned to enroll 450 patients and target 375 events to detect an improvement in median survival from 5 months to 7 months with hazard ratio 0.71 with 90% power. However, if the true hazard ratio is 0.77 which is worse than 0.71 but still clinically meaningful, the study only has 70% power. To mitigate the risk of being underpowered, one interim analysis was planned when 50% death events were observed to increase both the planned events and sample size by 50% if the interim results fell into the promising zone. Early stopping for efficacy was also possible which was based on O'Brien-Fleming efficacy boundary derived from the Lan and DeMets (Lan and De Mets 1983) [12] error spending function. If the trial did not cross the efficacy boundary at interim time, the interim results were partitioned into the following zones based on conditional power:

- Futility: CP < 5%.
- Favorable zone: CP > = 90%.
- Promising zone: 30 < = CP < 90%.
- Unfavorable zone: CP < 30%.

If the conditional power was below 5%, the study might be stopped for futility. If the trial fell into the favorable zone or unfavorable zone, the plan was to continue

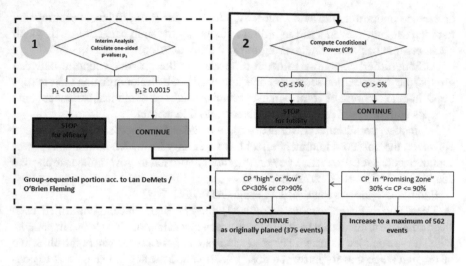

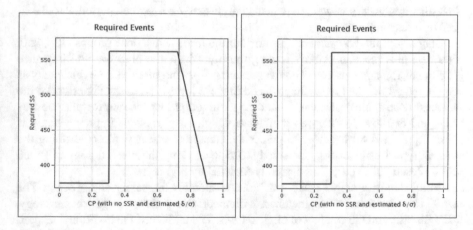

Fig. 4 Interim Decision Tree

Fig. 5 Sample size/events adaptation rule

the study as initially planned to enroll 450 patients and observe 375 events. If the trial fell into the promising zone, it was planned to enroll 676 patients and observe 562 events. The interim decision tree is depicted by Fig. 4.

At the time of the interim analysis, the SDAC prepared the report based on the interim analysis plan. After reviewing the interim results, the DMC made the recommendation to increase the sample size/events according to the pre-specified adaptation plan. The promising zone as originally intended would increase the number of events by targeting a conditional power of 90% and increase sample size proportionally to events. Such rule is shown by the left panel in Fig. 5 which could potentially allow backward calculation of the treatment effect observed at interim and hence undermine the study integrity. The VALOR trial ultimately employed a flat increase rule in events and sample size shown by the right panel of Fig. 5, which could prevent backward engineering of the treatment effect.

Although there was no significant difference in the primary endpoint between groups, the pre-specified secondary analysis stratified by randomization factors suggests that the addition of vosaroxin to cytarabine might be of clinical benefit to some patients with relapsed or refractory acute myeloid leukemia.

In practice, conditional power is often evaluated at the observed treatment effect. However, the observed treatment effect has variability for which conditional power cannot account for. Predictive power is another approach to govern the sample size adaptation (Mehta et al. 2022) [9].

Example: Seamless Phase II/III Design-ADVENT Trial:

The ADVENT trial was a randomized, double-blind, placebo-controlled two-stage adaptive clinical trial to assess the efficacy and safety of Crofelemer in patients with HIV-associated diarrhea. The study consisted of two stages where the objective of the first stage was to perform a dose selection and the second stage was to confirm the efficacy of the selected dose compared to placebo. The primary endpoint was control of watery bowel movements over a 4-week period. A patient who had less than two watery bowel movements per week over a 4-week period was classified as a responder.

Stage 1 enrolled 50 subjects per arm for the three active dose groups (125, 250, 500 mg) plus placebo. After selecting the right dose, the plan was to randomize additional 150 subjects equally to the selected dose and placebo. The final analysis combined all the data from Stage 1 and Stage 2 to test the treatment effect of the selected dose against placebo. Under the assumption that the rate for placebo was 35% and two low doses have no effect and high dose has 20% improvement, this design provided 80% power to detect the treatment effect while controlling the overall type 1 error rate at one-sided 0.025 level using the methodology in Posch et al. (Posch 2005) [10]. The design is depicted by Fig. 6.

In this study, a single DMC was used to monitor the safety and efficacy. The DMC included three voting members with medical experience in gastroenterology, HIV disease, and the conduct of clinical trials. The DMC had the traditional responsibility to monitor the ongoing safety by examining unblinded AE and SAE data. In addition to monitoring the safety of the trial participants, the other major responsi-

Fig. 6 Two-stage adaptive design

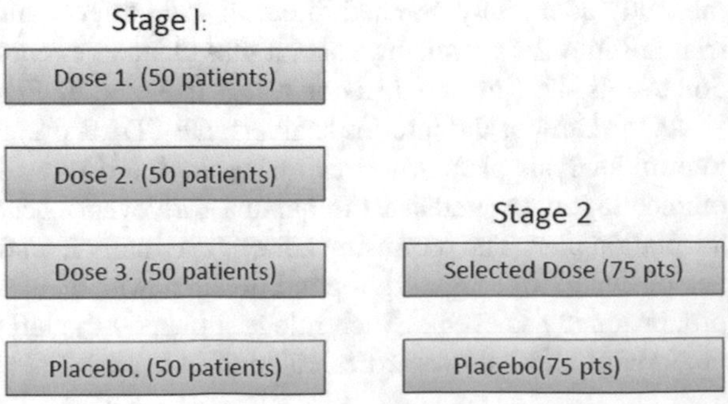

Stage I:

| Dose 1. (50 patients) |

| Dose 2. (50 patients) |

Stage 2

| Dose 3. (50 patients) | | Selected Dose (75 pts) |

| Placebo. (50 patients) | | Placebo(75 pts) |

bility was to implement the pre-specified dose selection criteria. The selected dose was revealed only to those personnel required to prepare and ship the study drug for Stage 2. The study also had a consulting statistician as a nonvoting member of the DMC. The consulting statistician played a dual role: (1) perform interim analysis according to the interim statistical analysis plan, (2) explaining the finer aspects of the adaptive design.

The dose selection criteria were as follows:

1. Assuming there are no safety issues, the Crofelemer dose selected for Stage 2 will be one for which the primary efficacy variable is at least 2% greater than the other Crofelemer treatments.
2. If two or three treatments groups are less than 2% of each other, and there no safety issues, the lowest of these doses will be selected for Stage 2.

Figure 7 shows the timing for the major milestones and events for the ADVENT trial. Stage 1 of the trial enrolled 194 subjects with 44 subjects for 125 mg, 54 for 250 mg, 46 for 500 mg, and 50 for placebo. The enrollment was paused until all Stage 1 subjects completed the placebo-controlled treatment period, the interim analysis and decision for Stage 2 were completed. The time point at which the last of the 194 subjects had completed the 4-week treatment period marked the start of the interim analysis period. After data were cleaned, a data cut was taken, and the cleaned interim data were sent to the independent statistician from the CRO. The independent statistician then compiled the necessary efficacy and safety tables and listings, prepared an electronic copy of the interim analysis report. A DMC meeting was then convened. The second column in Table 3 shows the response rates for each arm. Based on the assessment of efficacy and safety and the dose selection criteria outlined in the DMC charter, the lowest dose of 125 mg Crofelemer was recommended for Stage 2 by the DMC. Figure 8 shows the information flow for the

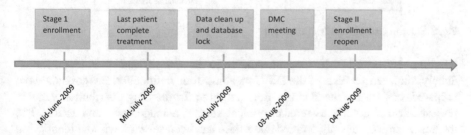

Fig. 7 Milestones and events

Table 3 Results for primary efficacy endpoint

Dose	Stage 1	Stage 2
Placebo	1/50 (2%)	10/88 (11.4%)
125 mg	9/44 (20.5%)	15/92 (16.3%)
250 mg	5/54 (9.3%)	
500 mg	9/46 (19.6%)	

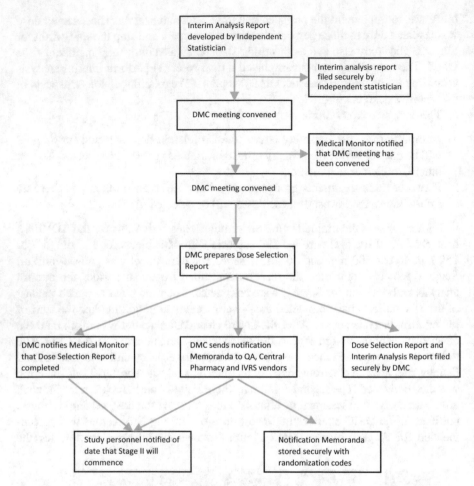

Fig. 8 Interim analysis and DMC meeting

interim analysis and DMC recommendation to the sponsor and study team. Immediately upon termination of the DMC meeting, four notification memoranda were
prepared, one for the medical monitor, and three for the drug distribution vendors
responsible for quality assurance, clinical supply management, and IVRS. The
medical monitor was only notified that a dose had been selected without identifying
the dose. The three drug distribution vendors were given the identity of the selected
dose. This marked the end of the interim analysis period. Enrollment was resumed
to the selected dose and placebo. The duration of interim analysis period was kept
at just 8 weeks.

To protect the study integrity, very strict procedures were applied to prevent from
the interim results leakage. All analyses were prepared by the SDAC who had no
direct involvement in the study conduct such as site monitoring and data management. The statistical software files used to prepare data tables, listings, figures, and
the interim analysis report were stored securely such that neither the sponsor nor
contract research organization (CRO) could access them. The randomization code

was stored on a computer that sponsor could not access, and the selected dose was revealed only to those personnel required to prepare and distribute the study drug for Stage 2.

In the traditional drug development paradigm, sponsors are responsible for dose selection. But for trials with seamless two stage adaptive design, sponsors are typically blinded to the selected dose to prevent potential operational bias. DMC review interim efficacy and safety data to make dose recommendations on behalf of the sponsors. Although the dose selection rule was laid out in the DMC charter, the rule was just a guidance. If there are safety concerns or the dose response pattern was beyond expectation, it might be very challenging to make the dose recommendation. If such scenarios happen, DMC might reach out to the Sponsor Liaison to discuss. The Sponsor Liaison personnel should be identified prior to interim analysis. Furthermore, if there is a plan to do accelerated submission based on the interim results, a separate team should be identified a priori to prepare the submission package. Since this team would have been unblinded in preparing the submission package, they cannot be involved in the study conduct going forward.

The other special aspect about the ADVENT trial was that the enrollment was halted after Stage 1 enrollment was completed so that the results of the first cohort of patients could be completed and the selected dose could be given to all the patients in the second cohort. This was possible because the study endpoint was observed fairly quickly, in four weeks. It is usually not feasible to pause the enrollment during the interim analysis because the sites that are enrolling patients, if shut down, would be difficult to re-open for the same trial. This is more common for enrollment to continue while the data from the first cohort are being analyzed and presented to the DMC. In this case, some of the patients who are enrolled into the second cohort while the interim analysis of the first cohort is being conducted will be randomized doses that will eventually be dropped and thus will not contribute to the final analysis. Therefore, it is very important to minimize the interim analysis time to avoid too much overrun.

It is also very important to keep the identity of the selected dose blinded to all parties except the ones who need to handle drug supply to avoid the potential operational bias. This is the case in trials for which it might take long time for the treatments to reach the clinically meaningful effects. Dose selection based on the primary endpoint might not be feasible in such trials since enrollment would have been completed by the time all Stage 1 patients complete the assessment for the primary endpoint. Dose selection is often based on biomarker or surrogate endpoints in trials with survival or longitudinal outcome. For example, Carreras et al. (2015) [11] discussed adaptive seamless designs with interim treatment selection with survival data in oncology trials. In such trials, overall survival takes longer time to accrual enough events to differentiate the different doses. Progression survival or tumor response rate might be used as surrogate endpoint for dose selection. At the time of interim analysis, patients who are enrolled in Stage 1 are still being followed for the primary endpoint. These patients who are still being followed for the primary endpoints after interim analysis should not have the knowledge about the selected dose since knowing the identity of the selected dose could lead to unexpected dropouts prematurely and ultimately impact the statistical power of the study and/or introduce bias.

Example: Adaptive Enrichment Design-TAPPAS Trial.

TAPPAS was a multinational, multicenter, open-label, parallel-group, phase 3 randomized clinical trial of TRC105 plus Pazopanib versus Pazopanib in patients with cutaneous and non-cutaneous advanced angiosarcoma. The primary endpoint was progression-free survival. The key secondary endpoint was overall survival.

There were a few factors which contributed to the decision for utilizing the adaptive enrichment design for TAPPAS trial. First, there was some indication that TRC105 might work better for the cutaneous subgroup but there was not enough data to preclude the scenario that TRC105 works for the full population. Due to the ultra-orphan status of the disease and the paucity of reliable prior data on PFS or OS, an adaptive design with an unblinded interim analysis to modify the study in the following two aspects was implemented: sample size re-estimation and population enrichment. The study was initially designed to enroll a total of 190 subjects and collect 95 PFS events. This sample size provides 83% power at one-sided significance level 0.025 to detect an improvement in median PFS from 4 months to 7.27 months which corresponds to a hazard ratio of 0.55. To control the family-wise error rate, a novel approach proposed by Jenkins (Jenkins 2011) [13] was used which split the patients and PFS events into two cohorts with 120 patients for Cohort 1 and 70 patients for Cohort 2. Such approach allows DMC to fully use all available data at the time of interim analysis and use clinical judgment to make recommendation to the sponsor for the second stage of the study. The plan was to collect 60 PFS events from Cohort 1 and 30 events from Cohort 2. An unblinded interim analysis was planned when 40 PFS events were observed from Cohort 1 or 30 days after 120 patients enrolled. Figure 9 shows adaptations for Stage 2 after DMC reviewing all the data from Cohort 1 patients.

After 40 PFS events were observed, conditional power for the full population and cutaneous subpopulation would be computed, denoted by CF_F and CF_S respectively. The interim results of the trial were classified into four zones based on conditional power in the full population and subpopulation. If $CF_F > 95\%$, the trial was considered falling into the favorable zone where the trial will continue as planned to enroll 70 patients and collect 35 events for Cohort 2. If $30\% < CF_F < 95\%$, the trial was considered falling into the promising zone where sample size increase will be triggered. In the promising zone, the plan was to enroll 220 patients and collect 110 events for Cohort 2. If $CF_F < 30\%$ and $CF_S > 50\%$, the trial was considered falling into the enrichment zone where the enrollment for Cohort 2 would be restricted to the cutaneous subgroup to enroll 160 patients and collect 110 events. The other zone was unfavorable zone which was defined as $CF_F < 30\%$ and $CF_S < 50\%$. In the unfavorable zone, the trial would continue as planned to enroll 70 patients and collect 35 PFS events. Note that Cohort 1 was not modified no matter what the interim results were and only Cohort 2 was adapted if the trial fell into the promising zone or enrichment zone.

Conditional power is the probability of achieving statistical significance assuming the observed trend continues after interim. Let Z_1^F denote the Wald statistic (i.e., standardized log rank statistic) at the interim analysis comparing treatment to control in the full population, $d_2^F = 95$ denote the total number of events initially

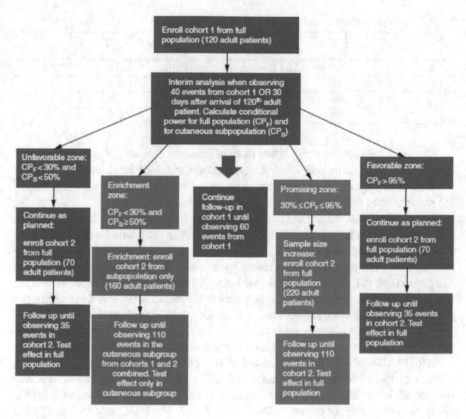

Fig. 9 Interim analysis flowchart

planned for the trial, and $d_1^F = 40$ denote the number of events at the interim analysis. The conditional power could be computed by the following formula assuming a 1-sided significance level of α.

$$CP_F = \Phi\left(\Phi^{-1}(\alpha)\sqrt{\frac{d_2^F}{d_2^F - d_1^F}} - \left[\sqrt{\frac{d_1^F}{d_2^F - d_1^F}} + \sqrt{\frac{d_2^F - d_1^F}{d_1^F}}\right]Z_1^F\right)$$

Similarly, let d_1^S be the number of events for cutaneous subgroup at interim time, $d_2^S = 110$ be the total number of events for cutaneous subgroup at final analysis in case of enrichment, Z_1^S be the Wald statistic for cutaneous group at the interim analysis. Then the conditional power for the cutaneous subgroup is given by

$$CP_S = \Phi\left(\Phi^{-1}(\alpha)\sqrt{\frac{d_2^S}{d_2^S - d_1^S}} - \left[\sqrt{\frac{d_1^S}{d_2^S - d_1^S}} + \sqrt{\frac{d_2^S - d_1^S}{d_1^S}}\right]Z_1^S\right)$$

Table 4 Example scenarios for interim results

Z_1^F	d_1^F	d_2^F	Z_1^S	d_1^S	d_2^S	CP_F	CP_S
−2.2	40	95	−1.5	18	110	97%	97%
−1.5	40	95	−0.8	18	110	68%	51%
−1	40	95	1.2	18	110	29%	86%
−1	40	95	−0.7	18	110	29%	40%

Where Φ denotes the cumulative distribution function of the standard normal distribution. This definition of conditional power assumes negative values of the Wald statistic indicate estimated hazard ratios below 1, i.e., outcomes favorable to the experimental arm.

Table 4 illustrates the possible scenarios for the interim results. The first scenario shows a scenario where the trial falls into the favorable zone with conditional power for the full population greater than 95%. The second scenario corresponds to the interim result falling into the promising zone where the conditional power for the full population is greater than 30% and below 95%. The third scenario shows that the conditional power for the full population is below 30%, but the conditional power for the cutaneous subpopulation is greater than 50%. The last scenario corresponds to the one where conditional power for full population is below 30% and the conditional power for the cutaneous subpopulation is below 50%.

In TAPPAS trial, the DMC consisted of two clinicians and one statistician. The DMC was tasked to provide oversight of safety and efficacy considerations and provide advice to the sponsor regarding actions the DMC deemed necessary for the continuing protection of patients enrolled in the trial. The DMC was also charged to monitor study design assumptions, determine whether the overall integrity and conduct of the study remained acceptable, advise on administrative changes to the protocol and make recommendations on procedures for data management and quality control. In addition, the DMC would also consider factors external to the study when relevant information became available, such as scientific or therapeutic developments that might have an impact on the safety of the participants or the ethics of the trial. Last, the DMC would periodically review accumulating safety data and make recommendations to the sponsor to continue, terminate, or modify the study based on the interim results. At the time of the planned interim analysis of efficacy, the DMC would make recommendations on the final sample size and/or population by the pre-specified adaptive design rules.

The responsibility of the sponsor included scheduling and facilitating DMC meetings, notifying the DMC of all changes to the protocol or study conduct, overseeing the collection and delivery of safety and efficacy data to the statistical analysis center, and communicating DMC recommendations to investigators, regulators, and study participants, as appropriate.

For the TAPPAS trial, the SDAC consisted of two statisticians as the members who received the unblinded interim data and randomization list from the corresponding vendors and prepared the tables listings and figures for efficacy data including conditional power to guide the interim decision. The two members of the

SDAC served as a non-voting independent statisticians supporting the DMC with respect to any issues that could arise concerning the design, monitoring, interim efficacy analysis or interim safety analysis for this adaptive trial. These two statisticians were present for both the Open and the Closed Sessions of the DMC meeting. The preparation of safety data for periodic DMC meetings was done by another group. The efficacy data summary included conditional power for the full population and cutaneous subpopulation, PFS based on blinded independent central review (BICR) and based on investigator, objective response rate (ORR), overall survival (OS) by treatment arm and overall. In addition, data presented to the DMC included plots by treatment group for the following non-overlapping categories: survival without progression and without treatment-related SAE, survival without progression with treatment-related SAE, survival with progression and without treatment-related SAE, survival with progression and with treatment-related SAE, death. The safety data include AE, SAE, labs, vital sign, ECG, ECOG, concomitant medications, and QoL (Quality of Life). The efficacy data was analyzed and validated by an independent statistician with experience and expertise in adaptive design.

The DMC meetings were planned to be held quarterly and no less frequently than 6 months via teleconference to review safety data. The interim analysis meeting for efficacy was face to face. The timing of data delivery were 2 weeks in advance of each DMC meeting. A meeting quorum required that all 3 voting members be present by phone or in person.

The recommendation from DMC to continue, modify or terminate the study would be forwarded to the sponsor within one day of each DMC meeting. Action items including requests for additional data review or statistical analysis would be forwarded to the sponsor within 3 weeks following the meeting. The possible recommendations from the DMC could be: (1) continue the study as planned, (2) make a minor modification in study conduct as specified below, (3) A further meeting is required to discuss analyses unavailable today and the analyses the DMC would like to review are specified below, (4) A major modification to the study should be considered as specified below, (5) Results of the interim analysis on determination of final sample size as specified below, (6) Study termination should be considered for the reasons specified below.

TAPPAS trial was an open label study where blinding was not feasible. Some efforts were taken to protect the study integrity. First, data management was handled by an external vendor instead of in house. A BICR was utilized to make adjudications on PFS. In oncology trials, if patient disease progresses, investigators often discontinue patients from the current treatment and move patients to alternative therapies. However, it is important for investigators to wait for the judgment from the central adjudication committee before moving patients to other treatment options since the primary efficacy endpoint is progression-free survival by central adjudication committee. If the disease only progresses based on investigator's judgment but not central committee assessment, the protocol may specify that those subjects be censored, and the ultimate events and study power might be reduced. At minimum, the potential treatment effect will be ameliorated if subjects take other treatments before it actually is clinically necessary (i.e., before disease progression has, in fact, occurred).

What Does the Paperwork from DMC Meetings Look Like?

David Kerr and Nand Kishore Rawat

Abstract This chapter gives guidance on the formal documentation that should be in place after each DMC meeting takes place. These include open session minutes, closed session minutes, recommendation form, and action item form. The process for drafting, reviewing, and finalizing are covered. Once complete, additional discussion with the Sponsor Liaison may be needed to document the sponsor response to the DMC recommendations.

Keywords Open minutes · Closed minutes · Recommendation form · Action item form · Sponsor Liaison

Formal documentation of Data Monitoring Committee (DMC) meetings and recommendations is critical. Minutes should be taken for all DMC meetings and meeting sessions. There typically are four documents created:

- Top-line recommendation form.
- Detailed recommendation/action item form.
- Open Session minutes.
- Closed Session minutes (one per protocol reviewed) including internal action items.

The minutes should include, at a high level, the information that was shared, key discussion points, and any decisions made. The minutes should maintain a factual and neutral tone and not ascribe any comment or decision to an individual unless it is so requested. The minutes needn't contain detailed information that is included in other study documents such as the Protocol or SAP or sponsor's Open Session

D. Kerr
Seattle, WA, USA
e-mail: david.kerr@cytel.com

N. K. Rawat (✉)
Lantheus Medical Imaging (United States), King of Prussia, PA, USA

N. K. Rawat, D. Kerr (eds.), *Data Monitoring Committees (DMCs)*,
https://doi.org/10.1007/978-3-031-28760-2_16

presentation. Rather, those documents may be referenced. There certainly is no need for a verbatim transcription or anything approaching that level of detail.

There typically is no reason to record the meeting – either to help minutes or for posterity. An experienced and sufficiently staffed Statistical Data Analysis Center (SDAC) will be able to take notes allowing for accurate and timely drafts of the minutes to be created. The reviewers of the minutes will fine-tune as needed prior to finalization. Under no circumstances should the meeting be recorded without the permission of the attendees.

Minutes typically will be signed by the Chair or acting Chair of the DMC and by the independent statistician at the SDAC who drafted them. Their signatures represent that the minutes accurately summarize the key elements of the occurrences and decisions from the meeting. Occasionally a representative from the Sponsor will sign the Open Session minutes.

The SDAC typically drafts minutes from both the Open and Closed Sessions of the meeting. If there is an Executive Session of the DMC at which the SDAC is excluded, the Chair or appointee should draft and archive the minutes. It typically is inefficient and awkward to have the sponsor draft minutes of the Open Session but have the SDAC draft minutes of the Closed Session.

A template DMC Recommendation form should be included as an appendix to the DMC Charter. It should include the standard possible recommendations (e.g., "Continue the study without modification", "Stop enrollment and/or dosing and/or study") as well as space for entering additional information or requests. As much as possible, the recommendation should be kept simple and brief. Additional information or requests the DMC has for the sponsor should be included in a separate Action Items document instead. The Action Items document could include a wide variety of topics, e.g., proposed next meeting date, request for additional information in the Open Session presentation in the future, recommendation for retraining of sites on certain aspect of the protocol, recommendation for new or updated tables or figures or listings in the future (unless these requested outputs are obviously motivated by by-arm differences and can be accommodated by the SDAC, in which case this recommendation would be in the Closed Session minutes for the SDAC to perform without notification of the Sponsor Liaison).

Attachments are not embedded in meeting minutes documents. Attachments within the minutes cause issues when documents are converted into final. PDF copies or employ digital signatures. This also reinforces that signatures are on the contents of the minutes documents themselves; the signatures do not apply to the contents of any documents presented during the meeting.

Separate Closed Minutes should be created for each protocol reviewed. If multiple protocols are reviewed during one Closed Session as part of a program-wide DMC review, a separate Closed Minutes document is written for each protocol. This allows minutes for one study to be provided upon study closeout even if other studies reviewed are ongoing. It eliminates partial information conveyed by page numbers if pages related to one protocol are redacted from a singular closed minutes document.

The level of detail should be appropriate. Enough detail should be included in the action items so that they are actionable by the appropriate party. Motivating factors for action items should not be included. Open minutes should include responses to questions raised by DMC members and need not include details available in the slide decks. Closed Minutes should include key discussion points that influenced the DMC's recommendation. The Closed Minutes need not include details of all tables reviewed by the DMC.

Action items may contain requests that could be informative if delivered to individuals involved in day-to-day study management. Action items should be delivered to the same individual who receives the overall study recommendation. That Sponsor Liaison may send the complete set of action items to the team or may need to follow up with a smaller group to address some of the action items presented there. Action items motivated (or likely to be interpreted externally as motivated) by by-arm differences that the SDAC can address independently of the sponsor should be documented in the Closed Session minutes and not included in the action items document provided to the Sponsor Liaison.

Attribution of comments in minutes should be minimal. Comments will not be attributed unless required to understand context, i.e., the DMC statistician explaining the implications of the monitoring boundaries.

The DMC makes recommendations as a collective unit. Thus, DMC signatures on Recommendation forms, Action Items, Open Minutes, and Closed Minutes should be limited to the DMC Chair. The independent statistician will typically co-sign the Open Minutes and Closed Minutes if that individual was the primary author of the documents.

The most efficient review process solicits feedback from the study team on the Open Minutes in parallel with the DMC reviewing the Open Minutes, Closed Minutes, and Action Items. Alternative processes include DMC Chair pre-review of all minutes or sponsor pre-review of the open minutes may reduce the exchange of comments but also increase effort associated with finalizing the documents and increase the time required to get final documents.

Review and finalization should take place in a timely way. The DMC recommendation should be complete and sent to the Sponsor Liaison within one business day of the meeting – particularly if the meeting was for an Interim Analysis, or there are any time-sensitive recommendations regarding patient safety. The meeting minutes and action items should be drafted within five business days so they are in the hands of reviewers before memories start to fade. Review would then be complete within five additional business days, with signature obtained quickly thereafter so that the meeting minutes and action items are complete within 2–2.5 weeks of the DMC meeting. If there are extensive revisions by a reviewer or contradictory edits made by different reviewers, then there might need to be further discussion or review before finalization.

The DMC recommendation should be delivered by the DMC, via SDAC, to the Sponsor Liaison. Ideally, the Sponsor Liaison is an individual with decision-making authority for the trial or who can forward DMC recommendations to those with decision-making authority. This individual generally has the scientific, medical, and clinical trial management experience to conduct and evaluate the trial. The Sponsor Liaison ideally is not involved with the day-to-day decision making of the trial. The Sponsor Liaison will be the recipient of DMC recommendations and be the first person to decide the course of action in responding to those DMC recommendations. After review, the Sponsor Liaison can share the high-level DMC recommendation (e.g., continue, stop) with the study team. In some situations, the Sponsor Liaison role is served by an independent executive committee of academic study leadership if such a group exists.

It is rare for the DMC to interact outside of the SDAC or Sponsor Liaison. However, there are occasions where the DMC has been asked to send confidential DMC materials directly to regulatory agency representatives. The DMCs do not talk directly to sites or Institutional Review Boards (IRBs) – typically simply the DMC recommendation letter is provided to these groups after each meeting. On occasion, DMCs overseeing the same treatment have been enabled to trade the minutes with each other after meetings. It is important that the Sponsor and external groups do not incorrectly interpret a recommendation of "Continue the study without modification." There have been instances where that was translated into a press release of "The DMC met and did not have any safety concerns." That is entirely incorrect. The DMC could have many safety concerns, but there simply were no alternative recommendations for the DMC to make at this time to the Sponsor Liaison. The DMC might be quietly planning with the SDAC for additional analyses and/or to meet again sooner than normal and be poised to make a non-trival recommendation at the next data review.

Examples of non-trivial recommendations could include the following:

- Met specified criteria for stopping for efficacy, but not enough safety data.
- Met specified criteria for futility on primary endpoint, but secondary endpoints actually look promising.
- Safety concerns – need to decide if

  - enrollment stopped in the entire study, or just a subset of arms.
  - treatment stopped in the entire study, or just a subset of arms.

- Update the protocol or Informed Consent Form (ICF).

  - Stop enrollment of a group now identified to be at high risk.
  - Mitigation plan to prevent events for those at high risk.

    - More frequent monitoring.
    - Dose reduction or interruption if patients are heading towards the safety event (e.g., if blood pressure starts to increase, or lab values change).
    - Update ICF with additional details if a new safety aspect is clearly identified during the course of the study.

It is common that there is back-and-forth with the Sponsor Liaison if the recommendation is other than 'continue the study without modification'. The DMC or a team led by the Sponsor Liaison might run outputs on subgroups or with sensitivity analyses before a final decision is made. For non-trivial recommendations, the

Sponsor Liaison should formally report back to the DMC on what the final decision within the Sponsor has been. There could be discussion between the Sponsor Liaison and the DMC to learn more about what led to the DMC recommendation. The Sponsor Liaison might propose a counter-proposal to the DMC. For example the DMC might have recommended treatment be stopped due to a concerning excess of Grade 4 neutropenia on new treatment, but the Sponsor Liaison proposes a dose management plan instead that reduces dose in subjects with Grade 3 neutropenia to hopefully prevent those patients from developing Grade 4 neutropenia. It is rare that the DMC is fundamentally in disagreement with the Sponsor decision – in particular if the decision impacts patient safety. It is preferable that an understanding can be reached between the DMC and Sponsor in such a case. It has been proposed that an outside mediator could be used, although this appears to be rare. If a DMC is fundamentally in disagreement and feels that patient safety or other ethics have been violated, the DMC has limited options. The DMC members – individually or *en masse* – can resign from the DMC. Eventually, the DMC meeting minutes will go to regulatory agencies, and the discontent will be discovered there. But the confidentiality documents signed by the DMC members prevent them from going public, even if there is this disagreement and even after resigning.

How Does the DMC Assess Risk-Benefit for Their Decision Making?

David Kerr and Nand Kishore Rawat

Abstract This chapter focuses on the DMC decision making, particularly when faced with difficult choices. These difficult choices could be due to data seen that reflects the integrity of the study, or in safety or efficacy domains. The concept of equipoise is introduced. The DMC has many options to help them in the case of difficult choices – either requesting more information and/or making non-trivial recommendations but not the ultimate level of recommending the stop of the study.

Keywords Risk-benefit · Ethics · Equipoise

The Data Monitoring Committee (DMC) members have a great responsibility to the current participants, the potential participants, and the global patient community. Most decisions made by the DMC will not have pure numeric guidelines in place. The DMC members – chosen for the experience and expertise – will have to weigh many factors to formulate an appropriate recommendation. They will have to weigh the risks seen, evidence of benefit, the clinical context, the patient population, etc. These decisions will focus on ethical considerations. It is a challenge, though, to evaluate ethics with the early, imprecise information provided to the DMC.

For example, a study might show early toxicity, but no hint of efficacy has yet emerged. If the indication is Type 2 diabetes, the context and subsequent recommendation likely would be quite different than if the indication is late-stage cancer. Patients with different conditions likely have a different threshold for the level of side effects that are acceptable. One would imagine that toxicity in a Type 2 diabetes treatment that is taken daily for years would have to be quite minimal, but toxicity for subjects with late-stage cancer would be much more acceptable.

D. Kerr
Seattle, WA, USA
e-mail: david.kerr@cytel.com

N. K. Rawat (✉)
Lantheus, King of Prussia, PA, USA

© The Author(s), under exclusive license to Springer Nature
Switzerland AG 2023
N. K. Rawat, D. Kerr (eds.), *Data Monitoring Committees (DMCs)*,
https://doi.org/10.1007/978-3-031-28760-2_17

Another example would be if a drug is already approved but is being used off-label frequently in another indication and is now being tested in that indication. If serious toxicity concerns emerge, the DMC might feel compelled to recommend the study continue to obtain unequivocal evidence of that serious toxicity to convince the clinical community. In this situation, the ethics of the global patient community (with millions of people) might take priority of the ethics of the patients currently in or potentially in the study.

The concept of equipoise is useful when thinking about clinical trials. A clinical trial is started where the sponsor has expectations on results, but really should enter unsure if the new treatment will have any impact. That is the equipoise – assuming all treatments will work equally well. The DMC members can ask themselves if equipoise is still maintained as the study continues.

To make it more personal, a DMC member might ask themselves:

- Would I accept my ill mother being enrolled on this study?
- Would I accept my ill mother being treated on the control arm of this study?
- Would I accept my ill mother being treated on the active arm of this study?

Despite all of the efforts of the sponsor, Statistical Data Analysis Center (SDAC), and DMC to have the needed material for review, some fears can arise to the DMC. Examples of these might be:

- There is a safety signal, but the numbers are too small to be certain. For example, in the domain of myocardial infarction, there might be six events on active vs. just two events on placebo. This is suggestive, but not definitive, of cardiac toxicity.
- There might be a safety signal, but the DMC is missing information on that domain completely. For example, in retrospect, the DMC wishes the protocol had mandated a full thorough QT study was conducted on each patient periodically, but this data is not available and will never be available.
- There is a potential safety concern, but alerting the sponsor to that will also have the effect of damaging study integrity. For example, there is now observed to be elevated heart rate on the active arm but no concerning excess in any cardiovascular adverse events yet. Should the DMC communicate this to the sponsor, if there is no specific other action to be taken? Or should the DMC, in consultation with the Sponsor Liaison, consider if action is needed to protect current and future patients with this knowledge of the elevated heart rate. Study integrity is best preserved if no information on by-arm results is communicated unless part of a DMC recommendation.

The ultimate decision of the DMC is whether to recommend stopping a study. The DMC could recommend stopping for various important reasons. These could include:

- Recommend a major change or stopping early for logistics.

  - The study is limping along – either with very low enrollment and/or few events in a time-to-event analysis. If the question the study is intended to answer will not be obtained for many years longer than originally expected, the DMC might recommend major change or even to stop the study.

- The study has a large number of subjects who have indicated withdrawal of consent or been lost to follow-up – perhaps differential by arm, or perhaps not. In either case, the DMC might consider if the study is no longer going to be interpretable at its end and therefore there is no purpose to the continuation of the study and might recommend major change or even to stop the study. Note that in many studies, subjects should continue on follow-up even if dosing is complete. It is important that tables clearly distinguish between discontinuation from treatment, and discontinuation from follow-up.
- The study has large number of subjects that enrolled that failed eligibility criteria and/or there are excessive number of subjects with major protocol deviations. In this case where the protocol is not being adhered to, the DMC might recommend major change or even to stop the study.

• Recommend a major change or stopping early for safety reason.

- This is the hardest decision of the DMC – and why there are experts on the DMC and not just computers.
- The DMC must weigh the totality of the data, and any informal efficacy data also provided.
- Three events of PML on the active arm versus none on the control arm might be sufficient to motivate a DMC to recommend major change or even to stop the study – but 60 vs. 20 cases of neutropenia might be totally expected and actually encouraging that the treatment is biologically active.
- Is the signal robust? Is it consistent across domains – e.g., do AEs and lab data correlate to tell the same story?
- Can AEs be combined for a more informative analysis – e.g., combine preferred terms of 'LDL increased', 'Lipids increased', 'Hyperlipidemia', and 'VLDL increased'?
- Is the signal known from pre-clinical results or completed clinical studies or as a class effect? Or is the signal novel – and therefore needs to be more compelling to be believed?
- Is the signal clinically relevant to the patient? Does it impact how the patient feels, function, and survives? If not, it may be of interest and perhaps some more minor action taken, but perhaps no major action needed at this time.
- Is the imbalance increasing from meeting to meeting?
- Is the safety concern offset by trends for positive efficacy?
- Are there baseline imbalances that could help explain additional safety events in one arm?
- Is there an imbalance in follow-up which means safety is biased towards reporting more events on one arm?
- Does the nature of the visit schedule (particularly in in an open-label study) have more assessment on one arm and therefore lead to bias towards more reporting of events?

– If this is a program-wide DMC, do the other studies show a similar trend or not?
– Is there a way this DMC can communicate with other DMCs also investigating the new treatment to investigate if similar trends across studies?

• Recommend a major change or stopping early for efficacy.

– Even with statistical guidelines (generally not rules), the totality of data should be reviewed. The DMC might look at secondary endpoints, or sensitivity results. The DMC might look at centrally adjudicated vs. investigator reported vs. best-case (adjudicated + events not yet adjudicated). The DMC might look at all data including recent data that is not fully cleaned, although the formal result is cut off at an earlier date and includes only pristine data.
– Ad hoc efficacy (outside of any specified guidelines) is controversial. The DMC should have discussion with the sponsor team during the kick-off meeting on this topic. If results on a clinically compelling endpoint (e.g., death) show statistically compelling results, even after accounting for the interim nature of the data, then many DMCs would feel obliged to report that to the Sponsor Liaison and recommend appropriate action such as treating all subjects with the better treatment (and possibly moving forward with regulatory approval).
– Ad hoc futility (outside of any specified guidelines) is also controversial. If there is no actual safety risk, some DMCs will recommend continuing the study. Some DMCs consider that a futile study is unethical to continue, as the subjects have entered the study expecting that their time and data will go towards obtaining useful information on a study that still has a hope for success. It is important that the sponsor explain the context of the study in the clinical program and regulatory environment to the DMC during the kick-off meeting and periodically during the course of the study. Most DMCs will have very little tolerance for safety risks in a study that appears to be statistically futile on primary (and clinically important secondary) endpoints.

• There are options available for DMC recommendation other than just 'go/no-go' when there are important concerns about safety – such as:

– Provide additional safety outputs – e.g., subgroups, SMQs.
– Provide additional efficacy outputs to see if those results counterbalance safety concern.
– Reinforce site training to be vigilant of specific safety issue.
– Change to protocol – mitigation plan (tighten eligibility criteria, safety management plan, dose management plan if precursor event seen).
– Change in meeting frequency (meeting more frequently, either formal meetings, or having ad hoc outputs focused on the domain of interest sent between meetings).
– Terminate enrollment but keep enrollment going for those already enrolled.
– Terminate enrollment and treatment, but keep follow-up going for those already enrolled.

What Are Some Examples?

David Kerr and Nand Kishore Rawat

Abstract This chapter provides a variety of vignettes culled from the author's experiences. These are not to be taken as actual results from historic studies but are reflective of the hundreds of DMC experiences seen. Example are from a wide range of clinical areas, and therefore wide range of endpoints and clinical context. The DMC focus on some examples is primarily safety-based, whereas others are primarily efficacy-based.

Keywords Data Monitoring Committee · DMC charter · Efficacy · Futility · Endpoint

The following vignettes are written to be illustrative of real-world situations, even though some specific details may have been changed for clarity and to protect the confidentiality of the studies.

Vignette #1: The Data Monitoring Committee (DMC) was asked to monitor two very large placebo-controlled studies of an investigational drug being conducted in parallel. The two patient populations were different but related – one being conducted in patients with Crohn's disease and the other in patients with ulcerative colitis. The DMC was asked to conduct quarterly safety reviews, as well as two formal interim analyses for each study. The purpose of the first formal interim analysis for each study was stopping for overwhelming efficacy. The purpose of the second formal interim analysis for each study was for stopping for overwhelming efficacy, but also for assessing futility. Particularly for assessing futility, the DMC was told to consider futility for each study separately, not referencing information from the other study. There was no binding futility rule, but rather a guideline that if the conditional power was <20% (under the original assumption of treatment effect under the alternative hypothesis), the DMC could consider recommend

D. Kerr
Seattle, WA, USA
e-mail: david.kerr@cytel.com

N. K. Rawat (✉)
Lantheus, King of Prussia, PA, USA

© The Author(s), under exclusive license to Springer Nature Switzerland AG 2023
N. K. Rawat, D. Kerr (eds.), *Data Monitoring Committees (DMCs)*,
https://doi.org/10.1007/978-3-031-28760-2_18

stopping for futility. This was a relatively weak, non-binding guideline. Although the DMC was told to assess each study independently, the DMC was to consider the impact of stopping one study while the other study continued, with no additional guidance given. In practice, the studies went on to completion (without a DMC recommendation to stop early for either efficacy or futility), with neither study meeting statistical criteria on the primary endpoint at the conclusion. One study (which completed first, about 4 months ahead of the second) had a point estimate in favor of the treatment and some secondary endpoints which trended favorably, while the second study had no observed effect. In retrospect, from the sponsor's perspective, the DMC charter could have used some stricter futility rules, although the DMC would have likely balked at a having a binding rule in the charter. The DMC felt the DMC charter was worded appropriately in providing some futility guidance from the sponsor's perspective while also giving the DMC sufficient freedom to assess futility and protect the scientific value of the studies.

Vignette #2: The DMC was overseeing a randomized open-label study in early-stage breast cancer. The study had co-primary endpoints of progression-free survival (PFS) and overall survial (OS). There were formal criteria in place to assess futility for PFS and efficacy for OS. The final number of events for PFS would occur earlier than the final number of events for OS. There was extensive discussion about the communication plan at the time of the final PFS analysis. The sponsor was considering unblinding themselves at the time of the final PFS analysis. The DMC advocated for the study team to remain blinded to ensure integrity of the study. This despite that it was open-label study, and that OS is not traditionally seen as easily biased. However, the DMC was worried that any knowledge of unblinded OS results within the study team or externally could impact how patients were handled in the study. The final decision was that if both PFS and OS were compelling at the time of the final PFS analysis, the study team would be told. Otherwise, only the PFS results would be told by the DMC to the Sponsor Liaison, and OS collection would continue without unblinding by the study team. If needed, the DMC could speak confidentially with regulatory authorities and speak in a limited fashion to the OS results if needed as part of an accelerated approval process based primarily on the PFS results.

The study played out with PFS results developing favorably as the study proceeded. However, OS results were neutral (fewer deaths due to disease progression were offset by more deaths due to infections). Unsurprisingly, criteria for futility for PFS and efficacy for OS were never crossed. At the time of the final PFS, the DMC informed the Sponsor Liaison only of the PFS results. The study continued, eventually finishing with a trend in improvement in OS, although not statistically compelling.

Vignette #3: This study was a randomized, double-blind study in acute myeloid leukemia. There were no formal interim analyses put into place. At the DMC Kick-off meeting, there was discussion that hematologic toxicity was expected – anemia, neutropenia, and thrombocytopenia. However, it was anticipated these would be relatively low grade/severity. Indeed, this toxicity was seen as the study data began to mature. However, it was more severe than expected. Grade 3 neutropenia was occuring in 23% of the patients on the active arm vs. just 10% on the control arm. Infections were reviewed, and there were higher rates of serious infections, but not as dramatic as in the hematologic parameters. The DMC noted that overall deaths were balanced and requested a Kaplan–Meier figure of time to death and a listing of Grade 3 neutropenia to understand the timing of those events. After review, it appeared the hematologic events were relatively transitory and occurred early after first treatment. The pattern of deaths shows an increase of deaths early on, but with the later results showing benefit. The DMC considered recommending a lower starting dose or a plan for dose reduction at the time of hematologic event, but ultimately decided that was not needed. The DMC was comfortable with the ethics of the study, and that there was still a reasonable likelihood of benefit to be demonstrated in OS once the initial toxicity had been survived.

Vignette #4: This was a randomized, double-blind study in colorectal cancer. No interim analysis reviews were planned. As the study matured, the DMC observed an excess in liver function test (LFT) abnormalities (ALT >3xULN and AST>3xULN) in the active arm. The DMC requested evaluation of drug-induced serious hepatotoxicity (eDISH) plots and patient listings of patients with elevated LFTs. The DMC recommended that the sites be particularly vigilant to sections of the protocol that already existed about monitoring of LFTs and subsequent dose reduction or withdrawal based on elevated values. The DMC also requested information on the primary endpoints – PFS as assessed by blinded independent central review (BICR) was not available, so investigator-proposed PFS was provided. As the study continued to mature, the elevated LFTs continued, and a case of laboratory Hy's Law occurred. The study had accrued over the half of the PFS events, but there was only a minimal improvement in PFS on the new treatment. Based on the observed concerning safety signal, and in the absence of any optimism on positive efficacy, the DMC recommended to stop the study. The rationale was based on both safety and futility concerns – although not based on statistical definition of futility.

Vignette #5: This study in pancreatic cancer had no formal interim analyses planned for efficacy. The sponsor had in place futility assessments for PFS, but no efficacy assessments during the study for PFS, and no interim analyses of OS at all, other than informally reviewing as part of the safety assessment. As the study developed, the OS results became progressively more and more impressive. A Kaplan–Meier figure was requested, and it showed graphically the strong and continued reduction in deaths on the new treatment. The DMC requested inferential statistics be provided. At what proved to be the DMC's final meeting (with the study being mature but still about a year from completion), the results showed that the OS results were both clinically compelling (hazard ratio of 0.37) and statistically compelling (p-value of 0.002). There were no new safety concerns beyond what was already known about the treatment going into the study. Although there were no specified decision criteria to evaluate for efficacy, the DMC decided to disclose the OS results to the Sponsor Liaison and recommend moving forward with regulatory approval and moving all subjects currently being treated with the control arm to instead be treated with the new treatment. This decision was partly based on ad hoc calculation of what a critical p-value would be if an O'Brien-Fleming boundary had been put into place for evaluating OS at this particular timing of the OS data. One could argue that this was data-driven and introduced bias – but the DMC felt the answer to the question of the study was known at that point, and it was unethical to continue subjects on the control arm to be treated in an inferior way, and important to move forward without delay to provide access to this new treatment to the larger patient population.

Vignette #6: This was a prophylaxis study for COVID-19 in 2020 and 2021. As would be expected, timelines were accelerated. This DMC met monthly at a specified date/time (e.g., 10 am ET on the first Tuesday of each month). The data arrived at the SDAC seven business days before each DMC review – allowing 4 business days for the SDAC to create outputs and 3 business days for DMC review. This worked because the incoming data and outgoing outputs were stable, and all parties involved carefully coordinated their timelines. The DMC met by teleconference for the first eight reviews. After that point, the DMC decided to do virtual reviews, with DMC members reporting thoughts directly by email to the DMC chair. That continued for six more reviews until the study finished. Of note for this DMC was the different approaches for providing information on whether a subject had had COVID-19 or not. Variations were based on AE data (i.e., AEs coded as "COVID-19" or "Asymptomatic COVID-19"), and information from a symptom questionnaire (which could use "strict" endpoint or "broad" endpoint), and also biologic information looking for positive reverse transcription-quantitative polymerase chain reaction (RT-qPCR). All information was useful to the DMC as they weighed evidence of efficacy against potential safety concerns of the new treatment.

Vignette #7: This study was relatively small and with short duration – just 120 patients, double-blind treatment throughout the first 12 weeks, with a 12-week endpoint (after which all subjects could be on active treatment and long-term followup). The endpoint was dichotomous – composed of whether subjects deteriorated and needed surgery prior to Week 12 OR if certain lab values had deteriorated at Week 12. There was a formal interim analysis planned at 50% of patients – 60 subjects reaching Week 12. The critical p-value was 0.001.

The results at the DMC interim analysis were striking – the numbers of subjects failing on placebo were strikingly higher than those on the new treatment – 17/30 failing on placebo vs. only 4/30 failing on the new treatment. This results in a p-value <0.001.

However, the DMC investigated the components of the composite outcome. The vast majority of cases (and imbalance) were due to the lab parameters, rather than subject actually requiring surgery. Surgery was split 3/30 vs. 0/30, whereas those with deteriorating lab results were split 14/30 vs. 4/30. The DMC wondered if efficacy primarily due to biomarker would be compelling to the clinical community – especially when based on just 60 subjects.

Safety data was reviewed – not just for the 60 who had reached Week 12, but also separately in an analysis that included all 90 that had entered the study so far. Nothing of particular concern was found.

There was discussion on how to handle the 30 subjects currently in the study, and the 30 subjects yet to be enrolled – and what the ethical obligations were to those subjects. However, the DMC decided it was still ethical to treat and enroll subjects on placebo for up to 12 weeks.

The DMC felt obliged to report to study leadership that the results had met the specified criteria. However, the DMC advocated that the study continue – given the small sample size so far, and that the endpoint results were dominated by the biomarker component, and that the study would naturally finish with 120 subjects reaching Week 12 within approximately a half year. After intense and rapid discussion between a small group of study leadership and the DMC (led by the DMC Chair), an agreement was reached to continue the study to its natural conclusion. The study results were released approximately a half year later with statistically significant results on the full 120 subjects (including more subjects who had surgery which continued to show a prominent trend).

References

[1]. US FDA. Guidance for clinical trial sponsors: establishment and operation of clinical trial data monitoring committees. Rockville: CBER/CDER/CDRH. US FDA; 2006.

[2]. DeMets DL, Fleming TR, Rockhold F, et al. Liability issues for data monitoring committee members. Clin Trials. 2004;1:525–31.

[3]. AICPA. Conceptual framework toolkit for independence, July 2022. https://us.aicpa.org/content/dam/aicpa/interestareas/professionalethics/resources/downloadabledocuments/toolkitsandaids/conceptual-framework-toolkit-for-independence-final.pdf

[4]. Wasserstein RL, Lazar NA. The ASA statement on p-values: context, process, and purpose. Am Stat. 2016;70(2):129–33. https://doi.org/10.1080/00031305.2016.1154108.

[5]. U.S. Food and Drug Administration. Guidance for industry. Adaptive Design Clinical Trials for Drugs and Biologics Guidance for Industry. 2019; https://www.fda.gov/regulatory-information/search-fda-guidance-documents/adaptive-design-clinical-trials-drugs-and-biologics-guidance-industry

[6]. Sanchez-Kam M, Gallo P, Loewy J, Menon S, Antonijevic Z, Christensen J, Chuang-Stein C, Laage T. A practical guide to data monitoring committees in adaptive trials. Ther Innov Regul Sci. 2014;48(3):316–26.

[7]. Antonijevic Z, Gallo P, Chuang-Stein C, Dragalin V, Loewy J, Menon S, Miller E, Morgan C, Sanchez M. Views on emerging issues pertaining to data monitoring committees for adaptive trials. Ther Innov Regul Sci. 2013;47:495–502.

[8]. Olshansky B, Bhatt DL, Miller M, Steg PG, Brinton EA, Jacobson TA, Ketchum SB, Doyle RT Jr, Juliano RA, Jiao L, Granowitz C, Tardif J, Mehta C, Mukherjee R, Ballantyne CM, Chung MK. REDUCE-IT INTERIM: accumulation of data across prespecified interim analyses to results. Eur Heart J Cardiovasc Pharmacother. 2021;7:e61–3.

[9]. Mehta C, Bhingare A, Liu L, Senchaudhuri P. Optimal adaptive promising zone designs. Stat Med. 2022;41(11):1950–70.

[10]. Posch M, Koenig F, Branson M, Brannath W, Dunger-Baldauf C, Bauer P. Testing and estimation in flexible group sequential designs with adaptive treatment selection. Stat Med. 2005;24(24):3697–714.

[11]. Carreras M, Gutjahr G, Brannath W. Adaptive seamless designs with interim treatment selection: a case study in oncology. Stat Med. 2015;34(8):1317–33.

[12]. Lan KKG, DeMets DL. Discrete sequential boundaries for clinical trials. Biometrika. 1983;70:659–63.

[13]. Jenkins M, Stone A, Jennison CJ. An adaptive seamless phase II/III design for oncology trials with subpopulation selection using correlated survival endpoints. Pharm Stat. 2010; https://doi.org/10.1002/pst.472.

N. K. Rawat, D. Kerr (eds.), *Data Monitoring Committees (DMCs)*, https://doi.org/10.1007/978-3-031-28760-2

Index

Printed in the United States
by Baker & Taylor Publisher Services